T0213437

SpringerBriefs in Mathematics

SpringerBriefs present concise summaries of cutting-edge research and practical applications across a wide spectrum of fields. Featuring compact volumes of 50 to 125 pages, the series covers a range of content from professional to academic. Briefs are characterized by fast, global electronic dissemination, standard publishing contracts, standardized manuscript preparation and formatting guidelines, and expedited production schedules.

Typical topics might include:

A timely report of state-of-the art techniques A bridge between new research results, as published in journal articles, and a contextual literature review A snapshot of a hot or emerging topic An in-depth case study A presentation of core concepts that students must understand in order to make independent contributions

SpringerBriefs in Mathematics showcases expositions in all areas of mathematics and applied mathematics. Manuscripts presenting new results or a single new result in a classical field, new field, or an emerging topic, applications, or bridges between new results and already published works, are encouraged. The series is intended for mathematicians and applied mathematicians. All works are peer-reviewed to meet the highest standards of scientific literature.

Titles from this series are indexed by Scopus, Web of Science, Mathematical Reviews, and zbMATH.

Alexander M. Blokh · Oleksandr M. Sharkovsky

Sharkovsky Ordering

Alexander M. Blokh
Department of Mathematics
University of Alabama at Birmingham
Birmingham, AL, USA

Oleksandr M. Sharkovsky
Institute of Mathematics
National Academy of Sciences of Ukraine
Kiev, Ukraine

ISSN 2191-8198   ISSN 2191-8201   (electronic)
SpringerBriefs in Mathematics
ISBN 978-3-030-99123-4   ISBN 978-3-030-99125-8   (eBook)
https://doi.org/10.1007/978-3-030-99125-8

Mathematics Subject Classification: 37E05, 37E15, 39A28, 37B40

This Springer imprint is published by the registered company Springer Nature Switzerland AG
The registered company address is: Gewerbestrasse 11, 6330 Cham, Switzerland

Preface

In 1964, a paper "Coexistence of cycles of a continuous mapping of the line into itself", by O. M. Sharkovsky, appeared in the Ukrainian Mathematical Journal, vol. 16 (1964), pp. 61–71. The following order was introduced among all positive integers:

$$3 \succ 5 \succ 7 \succ \ldots 2 \cdot 3 \succ 2 \cdot 5 \succ 2 \cdot 7 \succ \vdots 2^2 \succ 2 \succ 1$$

and the following theorem was proven: *if $m \succ n$ then for any continuous mapping f of the real line into itself the existence of a cycle of the map f of period n follows from the existence of a cycle of the map f of period m.*

This led to the inception of a new direction of research in the theory of dynamical systems called *combinatorial dynamics*. A number of new papers appeared in which the authors attempted to make this result more precise by taking into account the mutual location of points on the line, or by considering various types of trajectories on the plane other than cycles (e.g., *homoclinic trajectories*), or by considering certain types of maps in higher dimension. In particular, in 1979, a paper "On Sharkovsky's cycle coexistence order", by Peter Kloeden, appeared in Bull. Austral. Math. Soc., vol. 29 (1979), pp. 171–177; in this paper, perhaps, the expressions "Sharkovsky ordering" and "Sharkovsky theorem" were first introduced, and it was proven that a similar theorem and the same ordering hold for the so-called *triangular* maps that can be defined in any dimension.

This book is devoted to the above-quoted surprising mathematical result. As its scope is limited, we discuss only selected developments and direct attention of the interested reader to the book by L. Alseda, J. Llibre, and M. Misiurewicz "Combinational Dynamics and Entropy in Dimension One", 2nd edition, Advanced Series in Nonlinear Dynamics, **5** (2000), a true encyclopedia of the named subject, for a wider context and further references.

Most of the book was written in collaboration by the authors. Chapter 5 "Historical Remarks" is written by O. M. Sharkovsky. The book uses two different versions of the full name of O. M. Sharkovsky. Namely, his full name in Ukrainian is Oleksandr

Mykolayovych Sharkovsky (O. M. Sharkovsky), and its Russian transliteration is Alexander Nikolaevich Sharkovsky (A. N. Sharkovsky).

We are grateful to M. Yu. Matvijchuk, who was involved in writing Sects. 1.1–1.3 of Chap. 1 in 2013 before he moved to Canada.

To conclude, we would like to express our gratitude to the referees of the manuscript for useful remarks.

Birmingham, AL, USA   Alexander Blokh
Kyiv, Ukraine   Oleksandr Sharkovsky
October 2020

Contents

Chapter 1
Coexistence of Cycles for Continuous Interval Maps

1.1 Introduction

Given a map $f : X \to X$, a point $x \in X$ is said to be *periodic* if there exists an integer $n > 0$ such that points $x, f(x), \ldots, f^{n-1}(x)$ are pairwise distinct while $f^n(x) = x$; in that case, n is said to be the *(minimal) period* of x under f. The set $\{x, f(x), \ldots, f^{n-1}(x)\}$ is called the *(periodic) orbit (of x)*. We will interchangeably use equivalent terms *periodic orbit* and *cycle*. Our main interest in the present book is studying the coexistence of periodic points of various periods of self-mappings of a closed interval and of the real line $\mathbb{R}$.

In fact, it is easy to see that the results concerning coexistence of periods for self-mappings of $\mathbb{R}$ can be deduced from the results on coexistence of periods of self-mappings of a closed interval. Indeed, if a map $g : \mathbb{R} \to \mathbb{R}$ has a periodic orbit A of period n then we can construct the composition $r \circ g : \overline{A} \to \overline{A}$ where $\overline{A}$ is the smallest interval containing A and r is the natural retraction of the real line onto $\overline{A} = [u, v]$ (thus, r is the identity map on $[u, v]$, collapses $(\infty, u]$ to u, and collapses $[v, \infty)$ to v). One can see that periodic orbits of g are in fact periodic orbits of f; hence, the results on coexistence of periods that hold for f will hold for g.

Because of the above, from now on we consider continuous self-mappings of a closed interval (say, $[0, 1]$) to itself. Denote by $\mathrm{Per}(f)$ the set of all periodic points of f, and by $\mathrm{P}(f)$ the set of all periods of all periodic points of f.

Let us to introduce the *Sharkovsky order:*

$$3 \succ 5 \succ 7 \succ \cdots \succ 2 \cdot 3 \succ 2 \cdot 5 \succ 2 \cdot 7 \succ \ldots 2^2 \cdot 3 \succ 2^2 \cdot 5 \succ 2^2 \cdot 7 \succ \ldots 2^2 \succ 2 \succ 1.$$

The Sharkovsky theorem consists of two parts. In the first one, it is proven that for self-mappings of a closed interval the presence of a periodic orbit of a certain period m implies the presence of periodic points of all periods that are Sharkovsky-weaker than m. The second part shows that every possible initial segment of the Sharkovsky order is realized on a continuous interval map as the set of periods of its periodic

© The Author(s), under exclusive license to Springer Nature Switzerland AG 2022
A. M. Blokh and O. M. Sharkovsky, *Sharkovsky Ordering*,
SpringerBriefs in Mathematics, https://doi.org/10.1007/978-3-030-99125-8_1

points. From now on we will call these two parts the *Forcing Sh-Theorem* and the *Realization Sh-Theorem*; combined into one they form the *Sh-Theorem*.

Forcing Sh-Theorem *If* $g : [0, 1] \to [0, 1]$ *is continuous,* $m \succ n$ *and* $m \in P(g)$ *then* $n \in P(g)$.

Denote by $Sh(k)$ the set of all integers m with $k \succ m$ or $m = k$, and by $Sh(2^\infty)$ the set $\{1, 2, 4, 8, \dots\}$. In other words, sets $Sh(i)$ with $i \in \mathbb{N}$ or $i = 2^\infty$ are all initial segments of the Sh-order. By the Forcing Sh-Theorem, for a continuous interval map f the set $P(f)$ of periods of its periodic points coincides with the set $Sh(i)$ for some i. A natural question is, whether all possible initial segments of the Sh-order are realized as sets $P(f)$ for appropriately chosen maps f. The answer is given by Realization Sh-Theorem.

Realization Sh-Theorem *If* $k \in \mathbb{N} \cup 2^\infty$ *then there exists a continuous map* $f :$ $[0, 1] \to [0, 1]$ *such that* $P(f) = Sh(k)$.

For the sake of brevity in the rest of the book, we will also use terms the *Sh-order* instead of "the Sharkovsky order", *Sh-stronger* and *Sh-weaker* for "Sharkovsky-stronger" "Sharkovsky-weaker", resp., etc.

The original proof of Forcing Sh-Theorem was rather involved and technical. However, it was elementary in the sense that it did not use any of deep tools of dynamics, analysis, or other mathematical sciences. Later on, numerous attempts were made to simplify the proof (see [22], for instance). However, the proof was still cumbersome and hard to follow. The situation persisted until, around 1980, almost simultaneously several independent papers were published (see [6, 7, 14, 23]) where the authors proposed a new way of proving the theorem, which is called now the "standard" proof of the Sh-Theorem. Specifically, they introduced a directed graph associated to a given cycle of the map. Each loop in the graph corresponds to a periodic orbit whose period is a divisor of the length of the loop, and some additional assumptions of non-reducibility of the loop ensure that the period equals *exactly* the length of the loop. For proving the theorem, one needs to choose a periodic orbit, which is usually of the Sh-strongest period, and study its directed graph in order to show that there are loops of any required length in it.

Later on, modifications and simplifications of the standard proof appeared [1, 10]. One of the most clear and conceptual proofs of the Sh-Theorem was given by Burns and Hasselblatt, see [8]. We are going to present this proof below. Before doing that, let us mention other known proofs of Forcing Sh-Theorem. Bau-Sen Du succeeded in collecting simple proofs to suit different tastes [11]. Besides several versions of standard proofs, there are also proofs without involving directed graphs there. Reference [12] provides probably the shortest known proof of the theorem, and it does not use directed graphs.

There exist alternative proofs of the Sh-Theorem for some special classes of continuous maps. For instance, one can prove (see [15] and [9]) the theorem for unimodal maps, i.e., those with exactly one interior critical point, via use of kneading theory. See [19] for exposition of the kneading theory, though note that the paper was circulated a long time before its publishing, which allowed Jonker to use the kneading theory for proving the partial case of Sh-Theorem a decade earlier. Since most of results of the kneading theory extend from unimodal to multimodal case, it is our hope that the Sh-Theorem can be proved in a similar fashion for the class of piecewise-monotone maps. For bimodal case (i.e., for maps with two interior critical points), this was done in [17, 18]. For another proof in the unimodal case see [13], where the bifurcation theory is involved. Moreover, in [2], it was shown that for a quite general family of C^1-smooth unimodal maps the superstable cycles, i.e., those containing the unique turning point, appear according to the Sh-order.

In regard to Realization Sh-Theorem, the first proof appeared in [20] and [21]. In the first reference, which, by the way, contains the proof of the Forcing Sh-Theorem, a map is constructed; the map has a cycle of period n and no cycles of Sh-stronger periods than n, for any $n \in \mathbb{N}$. The second reference presents an example of a so-called 2^∞ map, i.e., a map having periods of every power of 2 and nothing else.

Let us outline the original proof; after that we will discuss the other proofs, which appeared later. First of all, the original proof was constructive and was accomplished in three steps. On the first step, for any odd $n > 1$, the formula for a map having period n, but having no odd periods strictly between 1 and n, was explicitly written. The second step is constructing examples of maps with the Sh-strongest period n, for any $n \in \mathbb{N}$. This was done inductively via use of "square root" construction. Starting from a map with Sh-strongest period $n \in \mathbb{N}$, this construction gives a map with Sh-strongest period $2n$.

Thus, using the first step as the induction base, one can create example maps with any Sh-strongest period. The third step is to construct a 2^∞ map. In order to do this, one takes a sequence of pairwise disjoint invariant intervals converging to a point. On the k-th invariant interval, one defines a 2^k map, i.e., a map having periods $1, 2, 2^2, \ldots, 2^k$ and only them. Such maps were constructed on the second step. What is left is to define the map on each complement interval linearly and in the limit point by continuity. It is quite straightforward to check that we obtained a 2^∞ map.

Alsedà, Llibre, and Misiurewicz in their book [1] presented a new elegant proof of the Realization Sh-Theorem, which is less technical and significantly shorter than the original one. This proof was "inspired by the proof of the Sharkovsky Theorem for unimodal maps", and, thus, all possible sets of periods can be found among unimodal maps. The main idea of the proof is to consider the full tent map $f(x) = 1 - |2x - 1|, x \in [0, 1]$, which has all periods. Then they cut its peak at level $h \in [0, 1]$, i.e., consider the family of truncated maps $f_h = \max\{f, h\}, h \in [0, 1]$. Varying h from 1 down to 0 one gets all possible beginnings of the Sh-order as sets of periods of f_h. We reproduce details of this argument below, with a slight modification in that we are cutting the tent map from below, rather than above.

1.2 Proof of Forcing Sh-Theorem

The version of the proof we present is close to the one from [8]. The proof is organized as follows. In the first part of the proof, we will show that presence of a minimal cycle of some period forces presence of cycles of all Sh-weaker periods. In the second part, we will show that all trajectories of Sh-greatest cycles are in fact minimal cycles.

The main tool used is a directed graph associated to a cycle. The loops in the graph correspond to other cycles.

1.2.1 Loops of Intervals Force Periodic Orbits

It is a very specific partial case of the Brouwer Theorem that, if $f(I) \subset I$ for a closed interval I, then f has a fixed point in I. This simple fact can be proven, however, directly from the Intermediate Value Theorem. The nice thing about dimension one is that the converse inclusion ensures existence of a fixed point, too.

Lemma 1.1 *If (a)$f(I) \supset I$ or (b)$f(I) \subset I$ for a closed interval I, then f has a fixed point in I.*

Proof Consider the map $g(x) = f(x) - x$. Clearly, zeros of g are fixed points of f. Suppose that f has no fixed points. Then g has no zeros. By the Intermediate Value Theorem, either (1) $g|_I$ is positive or (2) $g|_I$ is negative.

(1) g_I is positive, then in case (a) we have a contradiction as then the left endpoint of I does not belong to $f(I)$, and in case (b) we have a contradiction as then the image of the right endpoint of I does not belong to $f(I)$.

(2) g_I is negative, then in case (a) we have a contradiction as then the right endpoint of I does not belong to $f(I)$, and in case (b) we have a contradiction as then the image of the left endpoint of I does not belong to I. $\square$

From Lemma 1.1, we get the first part of the Sh-order as any interval map has a fixed point. Specifically, if an interval map f has a periodic orbit $\{x_1 < x_2 < \cdots < x_n\}$ of any period n, then one has $f[x_1, x_n] \supset [x_1, x_n]$.

Lemma 1.2 *If f has any period $n \geq 1$, then it has period 1.*

For further steps, we will need the *directed graphs generated by interval maps*. These tools rely upon the following technical lemmas.

Definition 1.1 For two closed intervals I and J, the notation $I \to J$ means that $f(I) \supset J$. A *string of intervals* if a finite string $L = \{I_0 \to I_1 \to \ldots$. A *loop of intervals* is a loop $L = \{I_0 \to I_1 \to \cdots \to I_{n-1} \to I_0\}$; it is called *non-repetitive* if it is not a repetition of a shorter loop of intervals. Let us say that p *follows L* if $f^n(p) = p$ and $f^i(p) \in I_{i \,(\text{mod}\, n)}$, $i \geq 0$. Also, L is said to be *period-forcing* if any periodic orbit that follows L must be of period n.

The next definition helps one describe mutual location of intervals.

Definition 1.2 Two intervals are said to be *almost disjoint* if they have disjoint interiors.

Observe that in the vast majority of cases loops of intervals consist of almost disjoint intervals, however in Definition 1.1 this is not required.

Lemma 1.3 *Let $L = \{I_0 \to I_1 \to \cdots \to I_{n-1} \to I_0\}$ be a loop of intervals. Then there is a point p that follows L.*

Proof We have $f^n(I_0) \supset I_0$. So, Lemma 1.1 ensures the existence of an f^n-fixed point in I_0. To find a periodic point that follows the loop of intervals, we need to be more careful. Let us write $I \rightarrowtail J$ if $f(I) = J$. Whenever $I \to J$, one can find a smaller subinterval $K \subset I$ such that $K \rightarrowtail J$. Indeed, it suffices to take a minimal K with the property $f(K) \supset J$. Starting with the loop $I_0 \to I_1 \to \ldots I_{n-1} \to I_0$, and applying the above observation successively to each arrow, we arrive at

$$K_0 \rightarrowtail K_1 \rightarrowtail \ldots K_{n-1} \rightarrowtail I_0.$$

Now, we can apply Lemma 1.1 to the interval K_0, because $K_0 \subset I_0 = f^n(K_0)$. This gives us an f^n-fixed point in K_0, which has to follow the loop $I_0 \to I_1 \to \cdots \to I_{n-1} \to I_0$. □

It can happen that the period of p in Lemma 1.3 is strictly less than n, the length of the loop of intervals. That might be inconvenient for us, since we are interested in showing the existence of cycles of specific periods. There are several ways to bypass this problem. For instance, if we know that the intervals I_i, $0 \le i < n$ in Lemma 1.3 are pairwise disjoint, then obviously the periodic point p must have period n. However, disjointness is a bit superfluous condition, which is hard to achieve. To introduce a more realistic condition, we need a few definitions.

Definition 1.3 If $P \subset \mathbb{R}$ is a finite set (e.g., a cycle of a map) then each open interval whose endpoints are adjacent points of P is said to be P-*basic*. If the finite set P is already given we may omit it from notation and talk about *basic* intervals.

The condition from Lemma 1.4 will be extensively used in what follows.

Lemma 1.4 *Let $\mathscr{O}$ be a periodic orbit of a continuous interval map f. Let $L = \{I_0 \to I_1 \to \cdots \to I_{n-1} \to I_0\}$ be a non-repetitive loop of $\mathscr{O}$-basic intervals such that no point of $\mathscr{O}$ follows L. Then L is period-forcing and any periodic point that follows L is of period n.*

Proof Let $f^n(p) = p$ and $f^i(p) \in I_i$, $0 \le i < n$. Then the period of p divides n. Also, since the loop L is followed by p, but is not followed by any point of $\mathscr{O}$, the point p does not belong to $\mathscr{O}$ (and, hence, none of its iterates does). Now, recall that interiors of $\mathscr{O}$-basic intervals are pairwise disjoint. So, if the cycle $p \to f(p) \to \cdots \to f^{n-1}(p) \to p$ were a repetition of a shorter cycle, then the loop of intervals $I_0 \to I_1 \to \cdots \to I_{n-1} \to I_0$ would be a repetition of shorter loop of intervals, a contradiction. □

Abusing the language we will say that P is an n-*orbit* if P is a cycle of period n, and that a map f *has period* k if it has a cycle of period k.

1.2.2 The Beginning of the Sh-order

Lemma 1.5 *If f has period $n > 1$, then it has period 2.*

Proof We follow [3]. Let $\mathcal{O}$ be an n-orbit, $n > 2$, and $I_1, \ldots, I_{n-1}$ be all $\mathcal{O}$-basic intervals. Since for each i there is $j \neq i$ with $I_i \to I_j$, we can create a string of $\mathcal{O}$-basic intervals in which any two consecutive intervals are distinct. Clearly, a string constructed in this fashion will yield a non-repetitive loop L of intervals (it suffices to follow the string and stop when the same interval appears in it for the second time) of length p, $1 < p \leq n - 1$; evidently, no point of $\mathcal{O}$ follows L as $\mathcal{O}$ has period greater than p. Repeating these arguments, we will find a point of period 2.   □

Lemma 1.5 gives us all information about the beginning of the Sh-order.

Lemma 1.6 *The beginning of the Sh-order looks as follows:*

$$\{\text{non-powers of } 2\} \succ \cdots \succ 2^n \succ 2^{n-1} \succ \cdots \succ 2^2 \succ 2 \succ 1.$$

Proof We claim that if f has period that is not power of 2, then it has period 2^n for any $n \geq 1$. Indeed, let f have a period that is not a power of 2, and let 2^m be any power of two. Then $f^{2^{m-1}}$ also has a period that is not a power of two. By Lemma 1.5 $f^{2^{m-1}}$ has period 2. Therefore f has period 2^m.

Similarly, if f has period 2^n, $n \geq 2$, then f has period 2^{n-1}. Indeed, if f has period 2^n, $n \geq 2$, then $f^{2^{n-2}}$ has period 4. By Lemma 1.5, then, $f^{2^{n-2}}$ has period 2. Hence, f has period 2^{n-1}.

All this implies the desired initial segment of the Sh-order.   □

Lemma 1.7 *If f has period that is not a power of 2, then it has period 2^n for any $n \geq 1$.*

Proof Let f have a period that is not a power of 2, and let 2^m be any power of two. Then $f^{2^{m-1}}$ also has a period that is not a power of two. By Lemma 1.5 $f^{2^{m-1}}$ has period 2. Therefore f has period 2^m.   □

In fact, Lemma 1.5 explains how different powers of 2 are compared to each other. Specifically, let f have period 2^n, $n \geq 2$. Then $f^{2^{n-2}}$ has period 4. By Lemma 1.5, then, $f^{2^{n-2}}$ has period 2. Hence, f has period 2^{n-1}. Summarizing these argument, we get the following description of beginning of Sh-order.

1.2.3 Three Implies Everything

Lemma 1.8 *If f has period 3, then it has period n for each $n \geq 1$.*

Proof The following proof first appeared in [16]. Let $\mathcal{O} = \{x_1 < x_2 < x_3\}$ be a 3-cycle. For definiteness assume that $x_1 \to x_2 \to x_3 \to x_1$; set $I_1 = [x_1, x_2]$ and

$I_2 = [x_2, x_3]$. Then $I_1 \to I_2$, $I_2 \to I_2$, $I_2 \to I_1$. Given $n \neq 3$, consider the loop $L = \{I_1 \to \underbrace{I_2 \to \cdots \to I_2}_{n-1} \to I_1\}$. Clearly, L is not followed by any point of $\mathcal{O}$. Therefore, by Lemma 1.4 f has period n. □

Having established first few partial results let us return to the proof of the whole Sh-Theorem.

1.2.4 Minimal Cycles Imply Sh-weaker Periods

In this subsection, we introduce the notions of a *Štefan cycle* and of a *simplest cycle*. We prove that, whenever an interval map has a simplest cycle of period N, it has all the periods that are Sh-weaker than N.

1.2.4.1 Štefan Cycles

We will first define *Štefan cycles*. In what follows when considering cycles of a map f we will denote the first point of a cycle by $p_0, x_0, \ldots$ and the like while setting $f^i(p_0) = p_i$, $f^i(x_0) = x_i$, etc.

Definition 1.4 A periodic orbit of an odd period n is called a *Štefan cycle* if $n \geq 3$ and the orbit can be described as

$$p_{n-1} < p_{n-3} < \cdots < p_4 < p_2 < p_0 = p_n < p_1 < p_3 < \cdots < \cdots < p_{n-4} < p_{n-2}$$

or

$$p_{n-2} < p_{n-4} < \cdots < p_3 < p_1 < p_0 = p_n < p_2 < p_4 < \cdots < \cdots < p_{n-3} < p_{n-1}.$$

Evidently, any cycle of period three is a Štefan cycle. If the period is five this is no longer the case. Indeed, using the same notation as in Definition 1.4 we see that if $p_4 < p_2 < p_0 = p_5 < p_1 < p_3$ or $p_3 < p_1 < p_0 = p_5 < p_2 < p_4$, then the cycle is a Štefan cycle while it is easy to see that there are other cycles of period five too (such as, e.g., $p_0 = p_5 < p_1 < p_2 < p_3 < p_4$).

Proposition 1.1 *If an interval map f has a Štefan cycle $\mathcal{O}$ of odd period n, then for each m which is Sh-weaker than n there is a period-forcing m-loop of $\mathcal{O}$-basic intervals. In particular, f has period m.*

Proof The statement is vacuously true for $n = 1$, so let $n \geq 3$. Let I_1 be the basic interval between p_0 and p_1 and I_j be the basic interval between p_{j-2} and p_j for $1 \leq j \leq n - 1$. Clearly, $I_1 \to I_1$ and $I_j \to I_{j+1}$ for $1 \leq j \leq n - 2$. Besides, $I_{n-1} \to I_1 \cup I_3 \cup \cdots \cup I_{n-2}$.

Now, it is easy to find the period-forcing loops we need.

- If $m = 1$, then the loop $I_1 \to I_1$ is obviously period-forcing.
- If $m < n$ even, then the loop $I_{n-1} \to I_{n-m} \to I_{n-m+1} \to \cdots \to I_{n-2} \to I_{n-1}$ is period-forcing by Lemma 1.4. The points of the cycle $\mathcal{O}$ cannot follow this loop, since they are of period $n > m$.
- If $m > n$, then the loop $I_{n-1} \to \underbrace{I_1 \to \cdots \to I_1}_{m-n+2} \to I_2 \to I_3 \to \cdots \to I_{n-2} \to$

I_{n-1} is period-forcing, again, by Lemma 1.4. The points of the cycle $\mathcal{O}$ cannot follow this loop because there are at least three successive I_1's and only two points of $\mathcal{O}$ are inside of I_1 (the endpoints of I_1).

The last claim of the proposition follows from Lemma 1.4. $\qquad\qquad\square$

The concept of *simplest cycle* is more general.

Definition 1.5 Let P be a cycle of period n. If n is odd then P is said to be *simplest* if and only if either $n = 1$, or P is a Štefan cycle. In general, if n is even then P is said to be *simplest* if and only if its left and right halves are swapped by f and both are simplest cycles for f^2.

In particular, any cycle of period 2 is simplest. However, for other cycles of periods 2^n, the picture is more complicated. For example, consider a cycle X of period 4. Denote by x_0 its leftmost point and, as before, set $f^i(x_0) = x_i$. Then x_1 cannot be the adjacent to x_0 from the right point of X, however x_2 must be such. Hence, we have that either $x_0 < x_2 < x_3 < x_4$ or $x_0 < x_2 < x_4 < x_3$. Thus, there are only two simplest periodic orbits of period 4; clearly, there are cycles of period 4 that are not simplest (such as, e.g., $x_0 < x_1 < x_2 < x_3$).

1.2.4.2   Induction: All Periods

Proposition 1.2 *If an interval map has a simplest cycle $\mathcal{O}$ of period n, then for each m that is Sh-weaker than n there is a period-forcing m-loop of $\mathcal{O}$-basic intervals. In particular f has period m.*

Proof We use induction on the power of 2 in the prime factorization of n. For an odd n, Proposition 1.1 serves as a base of induction. Let $n = 2k$, assume that statement is true for k, and prove it for n. If $n \succ m$ then either $m = 1$ (in which case the proof is trivial) or $m = 2l$ with $k \succ l$. Assume the latter. Then by the inductive hypothesis, there is a period-forcing loop of length l

$$J_0 \to J_1 \to \cdots \to J_{l-1} \to J_0$$

for f^2 in $[x_1, x_k]$. Let K_i be the convex hull of $f(J_i \cap \mathcal{O})$. Then we get a loop of length $2l = k$ for f

$$J_0 \to K_0 \to J_1 \to K_1 \to \cdots \to J_{l-1} \to K_{l-1} \to J_0.$$

We claim that this loop is period-forcing. Indeed, take a periodic point p that follows the loop $J_0 \to K_0 \to J_1 \to K_1 \to \cdots \to J_{l-1} \to K_{l-1} \to J_0$. This p follows the loop $J_0 \to J_1 \to \cdots \to J_{l-1} \to J_0$ with respect to f^2 and, thus, has to have period l. Since $\mathcal{O}$ is simplest, f swaps the sets $[x_1, x_k] \cap \mathcal{O}$ and $[x_{k+1}, x_{2k}] \cap \mathcal{O}$. Hence the point p has period m with respect to f. □

1.2.5 Orbits with Sh-strongest Periods Form Simplest Cycles

The structure of the Sh-ordering implies that for a great variety of sets A of integers there exists the Sh-strongest number in A. The exceptional sets are infinite collections of powers of 2. Thus, if the set of periods of cycles of a continuous interval map f is not an infinite collection of powers of 2, then there exists the Sh-strongest integer in A. In particular, if the set of periods of cycles of f is not an infinite collection of powers of 2, then there exists a cycle of Sh-strongest period. Denote this period by m. We will show that any cycle of f of period m is simplest.

The notion of *swapping*, a version of what is also often called *division*, will be of great help in the arguments. For the duration of Subsect. 1.2.5 we fix some notation. Namely, we consider a cycle $\mathcal{O}$ (or P) of period $m \geq 2$. Let x be the rightmost point of $\mathcal{O}$ with the property $f(x) > x$, and let y be the point of $\mathcal{O}$ immediately to the right of x. Evidently, $f(x) \geq y$ and $f(y) \leq x$. Denote $J_1 := [x, y]$, and for each $i > 1$ let J_i be the convex hull of $f(\mathcal{O} \cap J_{i-1})$. Then $J_1 \to J_1 \to J_2 \to \ldots$ and consequently $J_1 \subset J_2 \subset J_3 \ldots$. Next, denote $\mathcal{O}_i := \mathcal{O} \cap J_i, i \geq 1$. One can see that $\mathcal{O}_1 \subset \mathcal{O}_2 \subset \mathcal{O}_3 \ldots$ and, since there are no proper invariant sets inside any cycle, this increasing sequence of sets stabilizes only when it reaches the maximal possible set, i.e., when $\mathcal{O}_i = \mathcal{O}$.

Definition 1.6 Say that there is *swapping* on the set $\mathcal{O}_k$ if, for any $z \in \mathcal{O}_k$,

$$z \leq x \Longrightarrow f(z) \geq y,$$
$$z \geq y \Longrightarrow f(z) \leq x.$$

As an example, consider a cycle P of an even period $m = 2k$ that has swapping on P itself. Then all points of $P \cap (-\infty, x]$ map into $[y, \infty)$, and all points of $[y, \infty)$ map into $(\infty, x]$. Hence, there are exactly k points of P in $(\infty, x]$, exactly k points of P in $[y, \infty)$, and these two halves of P exchange their places under the action of f. Such cycles are said to *have division*. Observe also that if P is a cycle of an odd period then swapping is impossible on the entire P.

1.2.5.1 Base Case: Odd Periods

Let $\mathcal{O}$ be a cycle of period $m \geq 2$. Let x be the rightmost point of $\mathcal{O}$ with the property $f(x) > x$ and let y be the point of $\mathcal{O}$ immediately to the right of x. By definition, we have $f(x) \geq y$ and $f(y) \leq x$. Let us denote $J_1 := [x, y]$, and for each $i > 1$ let J_i be the convex hull of $f(\mathcal{O} \cap J_{i-1})$. Then $J_1 \to J_1 \to J_2 \to \dots$ and consequently $J_1 \subset J_2 \subset J_3 \dots$. Next, denote $\mathcal{O}_i := \mathcal{O} \cap J_i, i \geq 1$. One can see that $\mathcal{O}_1 \subset \mathcal{O}_2 \subset \mathcal{O}_3 \dots$ and, since there are no proper invariant sets inside the cycle, this increasing sequence stabilizes only when it reaches the maximal possible set, i.e., when $\mathcal{O}_i = \mathcal{O}$.

Lemma 1.9 (Swapping lemma) *Let k be such a number that f has no odd periods less than $k + 2$, except fixed points. Then we have swapping on $\mathcal{O}_k$.*

Proof On the contrary, suppose it is not true for z and some k. Take $K \subset J_k$ to be either $[z, x]$ or $[y, z]$ depending on whether $z < x$ or $z > y$. Since z is mapped to the same side of (x, y) while both x and y change the side under f, we have $K \to [x, y] = J_1$. Also, $J_{k-1} \to J_k \supset K$. Based on that, we get the loops of k and $k + 1$ intervals:

$$J_1 \to J_2 \to \dots \to J_{k-1} \to K \to J_1; \quad J_1 \to J_2 \to \dots \to J_{k-1} \to K \to J_1 \to J_1.$$

Now, we apply Lemma 1.3 to one of these loops of intervals that has odd length and get a point $p \in J_1$ of an odd period less than $k + 2$ and such that $f^{k-1}(p) \in K$. By hypothesis, p should be a fixed point. In particular, $f^{k-1}(p) = p \in K \cap J_1 \subset \{x, y\}$. However, none of the points x, y is fixed.   □

The Swapping Lemma has far-reaching consequences concerning dynamics of cycles of maps which do not have certain periods.

Definition 1.7 Let P be a cycle. If there are adjacent points $u < v$ of P such that for each $z \in (\infty, u] \cap P$ we have $f(z) > z$ and for each $z \in [v, \infty) \cap P$ we have $f(z) < z$ then P is called *convergent*. Otherwise P is called *divergent*.

We are ready to prove Lemma 1.10.

Lemma 1.10 *Suppose that the Sh-strongest period of a cycle of a map f is m. Then the following holds:*

1. *If m is even then, except for fixed points, every cycle of f has division.*
2. *If m is odd then every cycle P of period m is such that the set P_{m-2} consists of all points of P except for either its leftmost point or its rightmost point, and we have strict inclusions*

$$\{x, y\} = P_1 \subset P_2 \subset \dots \subset P_{m-2} \subset P_{m-1} = P,$$

where each set P_i consists of $i + 1$ points. Moreover, P is convergent.

Proof (1) By the Swapping Lemma every cycle P of f which is not a fixed point has swapping. By the remark right after Definition 1.6, P has division.

(2) Evidently, every cycle of period 3 is convergent. Assume that $m \geq 5$. Then for $k = m - 2$ the map has no non-fixed cycles of odd periods less than $k + 2$. Let P be

a cycle of period m. Then by the Swapping Lemma the set P_{m-2} has swapping. The construction of sets $P_1 = \{x, y\}$, $P_2, \ldots, P_{m-2}$ shows that on each step at least one extra point is added to the set. Hence there are at least $m - 1$ points in the set P_{m-2}. On the other hand, by the remark after Definition 1.6, one cannot have swapping on the entire P because P is odd-periodic. Hence P_{m-2} consists of exactly $m - 1$ points. Since by construction each set P_i equals the intersection of a certain interval and P it follows that P_{m-2} coincides with P with either its leftmost point or its rightmost point removed, and, furthermore, we have that strict inclusions

$$\{x, y\} = P_1 \subset P_2 \subset \cdots \subset P_{m-2} \subset P_{m-1} = P,$$

where each set P_i consists of $i + 1$ points. Suppose that P is divergent. Then there must exist a point $v \in P$ such that $f(v) < v$ and the set $P \cap (-\infty, v]$ contains at least two points. The latter and the remark in the end of the previous paragraph imply that $v \in P_{m-2}$ and hence, by the Swapping Lemma, we must have $f(v) \geq y > v$, a contradiction. □

Proposition 1.3 is the main part of the proof of the Forcing Sh-Theorem.

Proposition 1.3 *If the Sh-strongest period of f is odd, then any orbit of this period is a simplest cycle.*

Proof Let $\mathcal{O}$ be a periodic orbit of the Sh-strongest period m. One may assume $m \geq 3$. Let x, y, and $\mathcal{O}_i$ be as defined before Lemma 1.9. At least one of the inequalities $f(x) > y$ and $f(y) < x$ takes place, because otherwise $\{x, y\}$ would be a 2-cycle. Let us assume that $f(y) < x$; since by Lemma 1.10 the cycle $\mathcal{O}$ is divergent, we can make this assumption without loss of generality. By Lemma 1.10 we have strict inclusions

$$\{x, y\} = \mathcal{O}_1 \subset \mathcal{O}_2 \subset \cdots \subset \mathcal{O}_{m-2} \subset \mathcal{O}_{m-1} = \mathcal{O},$$

where $\mathcal{O}_i$ consists of $i + 1$ points for $0 < i < m$.

Thus, $\mathcal{O}_2$ consists of points x, y, $f(x)$, and $f(y)$ of which exactly two must be the same. Since by the assumption $f(y) < x < y$, then $f(x) = y$ and $\mathcal{O}_2 = \{x, y, f(y)\} = \{x, f(x), f^2(x)\}$. Inductively, we get $\mathcal{O}_i = \{x, f(x), f^2(x), \ldots, f^i(x)\}$ for $1 \leq i < m$. Since each $\mathcal{O}_i$ is an intersection of a connected set with $\mathcal{O}$, each iterate $f^i(x)$ is located immediately next to the set $\mathcal{O}_{i-1}$. From the swapping on $\mathcal{O}_{m-2}$, we deduce successively $f^{m-1}(x) < \cdots < f^4(x) < f^2(x) < x < f(x) < f^3(x) < \cdots < f^{m-2}(x)$. □

1.2.5.2 Induction: All Periods

Here we generalize Proposition 1.3 to orbits of any periods.

Proposition 1.4 *Any cycle $\mathcal{O}$ of period n, that is, the Sh-strongest for an interval map f, is simplest.*

Proof As before, we will use induction on the power of 2 in prime factorization of n. The base of induction (the case of an odd n) is Proposition 1.3 (the case $n = 1$ is trivial). Assume that $n = 2k$ and the claim holds for k. Consider a periodic orbit $\mathcal{O}$ of period $2k$. By Lemma 1.10, $\mathcal{O}$ has division. That is, each point from $\mathcal{O} \cap [\min \mathcal{O}, x]$ is mapped under f to $\mathcal{O} \cap [y, \max \mathcal{O}]$ and vice versa. So, both left and right halves of the orbit $\mathcal{O}$ are f^2-orbits of period k. Observe that k is the Sh-strongest for f^2, for $2k$ is the Sh-strongest for f. By the inductive hypothesis, both left and right halves of $\mathcal{O}$ are simplest cycles for f^2. So, $\mathcal{O}$ itself is a simplest cycle for f. $\square$

So far we have proved that every cycle of the Sh-strongest period is simplest and that every simplest cycle of period m forces the existence of cycles of all periods that are Sh-weaker than m. Thus, we have proved the Forcing Sh-Theorem for the case when the set of periods of f has Sh-strongest period. The only remaining case is when f has every power of 2 among its periods, but nothing else. This exceptional case is treated separately below.

Proposition 1.5 *If all periods of f are powers of 2, then all its orbits are simplest cycles.*

Proof By induction on k we prove that all cycles of periods below 2^k are simplest. The base case $k = 0$ is trivial, as usual. The inductive step from k to $k + 1$ follows from the Swapping Lemma 1.9. $\square$

1.3   Proof of Realization Sh-Theorem

Let $f(x) = 1 - |2x - 1|, x \in [0, 1]$ be the standard tent map, which has finite number of cycles of period n, for any $n \in \mathbb{N}$. The map f has period 3, namely, $\frac{2}{7} \to \frac{4}{7} \to \frac{6}{7} \to \frac{2}{7}$. Therefore, by Forcing Sh-Theorem, f has cycles of all periods. Thus, for each $n \in \mathbb{N}$ we take the n-cycle B_n with the maximal possible $\min B_n$, and let

$$f_n(x) = \begin{cases} [l] f(x) & \text{, if } x \in \overline{B_n}, \\ f(\max B_n) & \text{, if } x > \max B_n, \\ f(\min B_n) & \text{, if } x < \min B_n, \end{cases}$$

where $\overline{B_n}$ denotes the convex hull of B_n. For each $n \in \mathbb{N}$, the map f_n obviously has a cycle of period n, so by Forcing Sh-Theorem, $P(f_n) \supseteq \mathrm{Sh}(n)$. In fact, we claim that $P(f) = \mathrm{Sh}(n)$. If these were not so, then there would have been a cycle $C_m \subset \overline{B_n}$ of period $m \succ n$. But then by applying Forcing Sh-Theorem, we would find a cycle of period n inside $\overline{C_m} \subset \overline{B_n}$. This contradicts the choice of B_n. And this proves that $P(f_n) = \mathrm{Sh}(n)$ for each $n \in \mathbb{N}$.

To finish the proof, we need to present a function g with $P(g) = \mathrm{Sh}(2^\infty)$. Consider the sequences $a_n = \min B_n$, $b_n = \max B_n$, $n \geq 1$. By definition of B_n and Forcing Sh-Theorem, we have that $a_m < a_n < b_n < b_m$, whenever $m \succ n$. Then, we define $a = \lim_{k \to \infty} a_{2^k}$, $b = \lim_{k \to \infty} b_{2^k}$. It is easy to see that the function

$$g(x) = \begin{cases} [l] f(x) & , \text{if } x \in [a, b], \\ f(\max B_n) & , \text{if } x > b, \\ f(\min B_n) & , \text{if } x < a \end{cases}$$

satisfies $P(g) = \text{Sh}(2^\infty)$, as desired. $\square$

1.4   Stability of the Sh-ordering

Whenever a mathematical phenomenon is discovered, a natural question follows: is this phenomenon stable with respect to small perturbations of a map that exhibits it? The importance of the problem of stability of phenomena becomes even more evident if we take into account the fact that numerically we never deal with the precise formulas. Hence, a crucial tool such as numerical simulation of a dynamical system inevitably shows the behavior of a system only close to the studied one and not necessarily coinciding with it. Thus, we are running a chance to numerically discover/confirm a dynamical phenomenon only if the phenomenon in question is stable with respect to small perturbations.

A simple example here is the so-called *attracting periodic points* of smooth maps, i.e., points x of period n such that f^n at the point x has multiplier of absolute value less than 1 (here we do not even have to talk about interval maps). Dynamically it would imply that neighborhoods of x map inside themselves under f^n. This allows one to use appropriate topological results and conclude that then there must exist an f-periodic point of period n close to x. The phenomenon of having an attracting periodic points of given period is, therefore, stable with respect to small smooth perturbations. Evidently, this concerns smooth maps.

The phenomena in question may be related to the map as a whole, not only to its individual points and their dynamics. In this section, we consider the central phenomenon of the book, i.e., the phenomenon of having a specific set of periods by interval maps, from the standpoint of its stability. The setting here is slightly different from the one considered above. Indeed, the Forcing Sh-Theorem deals with *continuous* interval maps. Therefore, it is natural to consider how sets of periods of periodic points of continuous interval maps vary under small perturbations of maps in C^0-topology. This is exactly the problem posed and solved by Louis Block in his paper [5]. The main result of [5] is the following theorem.

Theorem 1.1 ([5]) *Let f be a continuous map of the interval I to itself. Suppose that there exists an f-periodic point of period N. Moreover, assume that m is the number that immediately follows N in the Sh-order. Then there exists a C^0-neighborhood U of f such that for every map $g \in U$ the set of periods of g contains $\text{Sh}(m)$.*

Thus, a continuous interval map f with a periodic point of period 3 has a C^0-neighborhood U such that if $g \in U$ then g has a periodic point of period 5. On the other hand, observe that in general it is not necessarily so that a C^0-perturbation of a map f with a periodic point of period n itself has a periodic point of period n.

We will need the following simple lemma.

Lemma 1.11 (Lemma 10 [4]) *Let f be a continuous interval map that has a 4-periodic orbit $B = \{b_1 < b_2 < b_3 < b_4\}$ such that $f(\{b_1, b_2\}) \neq \{b_3, b_4\}$. Then f has a periodic point of period 3.*

Proof Set $x = b_1$. There are several ways in which points of B can be mapped to one another. The assumptions of the lemma imply that, up to orientation, these are as follows:

(1) $x = f^4(x) < f(x) < f^2(x) < f^3(x)$.
(2) $x = f^4(x) < f(x) < f^3(x) < f^2(x)$.
(3) $x = f^4(x) < f^3(x) < f(x) < f^2(x)$.

Considering each case separately one can show that in each of them the map f must have a point of period 3. As a matter of example, consider case (1). Set $[x, f(x)] = I$, $[f(x), f^2(x)] = K$, $[f^2(x), f^3(x)] = L$. Then $f(I) \supset K$, $f(K) \supset L$, and $f(L) \supset I$. It follows that there exists a point y such that $y \in I$, $f(y) \in K$, $f^2(y) \in L$ and $f^3(y) = y$. Clearly, y is a desired periodic point. $\qquad\square$

Proof of Theorem 1.1. Consider a point y and an odd number $2n + 1$ such that

$$f^{2n-1}(y) < f^{2n-3}(y) < \cdots < f(y) < y < f^2(y) < \cdots < f^{2n}(y)$$

and $y < f^{2n+1}(y)$. We claim that then f has a point of period $2n + 1$. Indeed, $f(y) < y < f^2(y)$ implies that there exists a fixed point $a \in (f(y), y)$. On the other hand, $f(f^{2n-2}(y)) = f^{2n-1}(y) < a < y < f^{2n+1}(y)$ which implies that there exists a point $a' \in (f^{2n-2}(y), f^{2n}(y))$ with $f(a') = a$. Consider a loop $\mathscr{A}$ of intervals $I_j = [a, f^j(y)]$ for $j = 0, 1, \ldots, 2n - 1$, $I_{2n} = [a', f^{2n}(y)]$ after which we complete the loop by putting I_0 again after I_{2n}. It is easy to see that in this loop of intervals each previous interval has the f-image covering the next one. Hence, the loop $\mathscr{A}$ gives rise to a periodic point z such that $f^j(z) \in I_j$, $j = 0, 1, \ldots, 2n$ and $f^{2n+1}(z) = z$. The point z is not fixed because, say, $[f(y), a]$ and $[a', f^{2n}(y)]$ are disjoint. Hence, z is of an odd period k with $1 < k < 2n + 1$. By the Forcing Sh-Theorem this implies the desired. Moreover, since all the inequalities above are strict it follows that there exists a small C^0-neighborhood U of f such that all continuous maps $g \in U$ have a periodic point of period $2n + 1$.

Next we prove the desired claim in the case when $N = 2k + 1$ is an odd number. By the Forcing Sh-Theorem we may assume that f has no periodic orbits of period $3, 5, \ldots, N - 2$. By the results of Chap. 1 we may assume that there exists a periodic point y such that

$$f^{2n-1}(y) < f^{2n-3}(y) < \cdots < f(y) < y = f^{2n+1}(y) < f^2(y) < \cdots < f^{2n}(y).$$

Since $f(y) < y < f^2(y)$ then there exists $b \in (f(y), y)$ with $f(b) = y$. Since $f^2(b) = f(y) < b < f(b) = y$ then there exists $d \in (b, f(b))$ such that $f(d) = b$.

Thus, $f^2(d) = y$. It is easy to check that the point d satisfies the conditions from the above paragraph for the odd number $N + 2$. Hence, by the results of the previous paragraph show that there exists a C^0-neighborhood U of f such that all maps $g \in U$ have periodic points of period $N + 2$ as desired.

Now, if f has a periodic point of period $2^r(2l + 1)$ where $l \geq 1$ then f^{2^r} has a periodic point of period $2l + 1$. By the above it follows that for some C^0-neighborhood of f and all maps $g \in U$ the maps g^{2^r} are sufficiently close to f^{2^r} so that g^{2^r} has a periodic point of period $2l + 3$ and, therefore, the map g itself has a periodic point of period $2^r(2l + 3)$ as desired.

To complete the proof of Theorem 1.1, we need to show that a if f has a periodic point of period $2^n, n \geq 1$ then there exists a C^0-neighborhood of f such that any map $g \in C^0$ has a periodic point of period 2^{n-1}. The claim is obvious if $n = 1$, so let us assume that $n \geq 2$. In fact let us first assume that $n = 2$ and the map f has a periodic point x of period 4. By the above we may assume that f has no periodic points of period 3. Hence, by Lemma 1.11 we have that, if $B = \{b_1 < b_2 < b_3 < b_4\}$ is the orbit of x then $f(\{b_1, b_2\}) = \{b_3, b_4\}$.

We may assume that $b_1 = x$. For the sake of definiteness assume that $x < f^2(x) < f(x) < f^3(x)$ (all other cases are considered similarly). Choose the greatest $u \in [x, f^2(x)]$ such that $f(u) = f(x)$. Then choose the least $v \in [u, f^2(x)]$ such that $f(v) = f^3(x)$. Evidently, $f([u, v]) = [f(x), f^3(x)]$ while $f^2(u) = f^2(x) > u$ and $f^2(v) = x > v$. Choose a sufficiently small C^0-neighborhood W of f. Then for any map $g \in W$ we will have that $g([u, v]) \cap [u, v] = \emptyset$ while $g^2(u) > u$ and $g^2(v) < v$. It follows that there exists a point $z \in [u, v]$ such that $g^2(z) = z$ while $g(z) \neq z$, i.e., a g-periodic point of period 2. Thus, regardless of the type of periodic point of period 4 that the map f has all sufficiently C^0-close to f maps will have a point of period 2.

Suppose that f has a periodic point of period $2^n, n \geq 2$. Then $f^{2^{n-2}}$ has a periodic point of period 4. By the previous paragraph there exists a small C^0-neighborhood U of f such that any map $g \in U$ has the property that $g^{2^{n-2}}$ has a periodic point of period 2. Clearly this implies that g itself has a periodic point of period 2^{n-1} as desired. $\square$

1.5 Visualization of the Sh-ordering

On the bifurcation diagram for the map $x \mapsto \lambda x(1 - x)$ below, we can see of the appearance of cycles of the least ten periods $1, 2, 4, 8, 10, 6, 9, 7, 5, 3$ just in Sh-ordering.

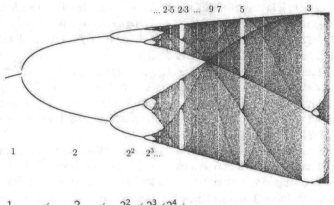

$$1 \quad \prec \quad 2 \quad \prec \quad 2^2 \prec 2^3 \prec 2^4 \prec \cdots$$

$$\cdots \prec 2^2 \cdot 7 \prec 2^2 \cdot 5 \prec 2^2 \cdot 3 \prec \cdots \prec 2 \cdot 7 \prec 2 \cdot 5 \prec 2 \cdot 3 \prec \cdots$$

$$\cdots \prec 9 \prec 7 \prec 5 \quad \prec \quad 3$$

References

1. Alsedà, L., Llibre, J., Misiurewicz, M.: Combinatonal dynamics and entropy in dimension one. In: Advanced Series in Nonlinear Dynamics, vol. 5, 2nd edn. World Scientific, River Edge, NJ (2000)
2. Arneodo, A., Ferrero, P., Tresser, C.: Sharkovskii's order for the appearance of superstable cycles in one-parameter families of simple real maps: an elementary proof. Comm. Pure Appl. Math. **37**, 13–17 (1984)
3. Barton, R., Burns, K.: A simple special case of Sharkovskii's theorem. Amer. Math. Monthly **107**, 932–933 (2000)
4. Block, L.: Simple periodic orbits of mappings of the interval. Trans. Amer. Math. Soc. **254**, 391–398 (1979)
5. Block, L.: Stability of periodic orbits in the theorem of Šarkovskii. Proc. Amer. Math. Soc. **81**, 333–336 (1981)
6. Block, L., Guckenheimer, J., Misiurewicz, M., Young, L.-S.: Periodic points and topological entropy of one dimensional maps. In: Global Theory of Dynamical System. Lecture Notes in Mathematics, vol. 819, pp. 18–34. Springer, Heidelberg (1980)
7. Burkart, U.: Interval mapping graphs and periodic points of continuous functions. J. Combin. Theory Ser. B **32**, 57–68 (1982)
8. Burns, K., Hasselblatt, B.: The Sharkovsky theorem: a natural direct proof. Amer. Math. Monthly **118**, 229–244 (2011)
9. Collet, P., Eckmann, J.-P.: Iterated maps on the interval as dynamical systems. Birkhäuser, Boston (1980)
10. Du, D.-S.: A simple proof of Sharkovsky's theorem. Amer. Math. Monthly **111**, 595–599 (2004)
11. Du, B.-S.: A collection of simple proofs of Sharkovsky's theorem. arXiv preprint http://arxiv. org/abs/math/0703592
12. Du, B.-S.: A simple proof of Sharkovsky's theorem revisited. Amer. Math. Monthly **114**, 152–155 (2007)
13. Guckenheimer, J.: On the bifurcation of the maps of the interval. Invent. Math. **39**, 165–178 (1977)
14. Ho, C.-W., Morris, C.: A graph theoretic proof of Sharkovsky's theorem on the periodic points of continuous functions. Pacific J. Math. **96**, 361–370 (1981)

15. Jonker, L.: Periodic orbits and kneading invariants. Proc. London Math. Soc. **39**, 428–450 (1979)
16. Li, T., Yorke, J.: Period three implies chaos. Amer. Math. Monthly **82**, 985–992 (1975)
17. Llibre, J., Mumbrú, P.: Periods and entropy for bimodal maps. Univ. Autònoma de Barcelona, preprint (1987)
18. Llibre, J., Mumbrú, P.: Renormalization and periodic structure for bimodal maps. In: European Conference on Iteration Theory (ECIT 87), pp. 253–262, World Scientific Publishing, Teaneck, NJ (1989)
19. Milnor, J., Thurston, W.: On iterated maps of the interval. Dynamical Systems. Lecture Notes in Mathematics, vol. 1342, pp. 465–563. Springer, Berlin (1988)
20. Sharkovsky, A.N.: Coexistence of cycles of a continuous mapping of the line into itself (in Russian). Ukrain. Math. Zh. **16**, 61–71 (1964)
21. Sharkovsky, A.N.: On cycles and the structure of a continuous mapping (in Russian). Ukrain. Math. Zh. **17**, 104–111 (1965)
22. Štefan, P.: A theorem of Šarkovskii on the existence of periodic orbits of continuous endomorphisms of the real line. Comm. Math. Phys. **54**, 237–248 (1977)
23. Straffin, P.D., Jr.: Periodic points of continuous functions. Math. Mag. **51**, 99–105 (1978)

Chapter 2
Combinatorial Dynamics on the Interval

2.1 Introduction

The Sh-theorem started a new field in the dynamical systems theory that can be appropriately called *(One-Dimensional) Combinatorial Dynamics*. In our presentation of basic results from the field, we basically follow notation and terminology from [2] and [31].

While the Sh-theorem is stated in the language of a specific ordering among the periods of cycles of an interval map, in reality, it solves the problem of fully describing all possible sets of periods of an interval map. Thus, a cycle is labeled by its period viewed as the type of the cycle, and we describe all possible sets of types of cycles of interval maps. Therefore, one direction of One-Dimensional Combinatorial Dynamics is to describe possible sets of types of periodic orbits of one-dimensional maps. This can be done by describing special ordering among types of cycles (so-called *forcing* relation) and then using it in the same way as the Sh-theorem is used for the full characterization of all possible sets of periods of interval maps. The core of this chapter is devoted to discussing, for interval maps, forcing relations among periodic orbits on the interval and their types, and, accordingly, describing possible sets of types of periodic orbits of continuous interval maps.

We shall continue using the techniques of cycles (loops) of intervals introduced earlier in Chap. 1. Because this tool is of the utmost importance in studying periodic points of interval maps, and for the convenience of the reader, we would like to recall the definitions. An interval J f-covers an interval K (we write then $J \to K$) if $K \subset f(J)$. If we have a *loop* of intervals $J_0 \to J_1 \to \cdots \to J_{n-1} \to J_0$ then there is a periodic point x such that $f^i(x) \in J_i$ for $i = 0, 1, \ldots, n-1$ and $f^n(x) = x$. We will say that the orbit of x is *associated* to the loop. Any piece $J_i \to \cdots \to J_j$ (or $J_j \to \cdots \to J_0 \to \cdots \to J_i$) of the loop will be called a *block (of the corresponding loop)*.

© The Author(s), under exclusive license to Springer Nature Switzerland AG 2022
A. M. Blokh and O. M. Sharkovsky, *Sharkovsky Ordering*,
SpringerBriefs in Mathematics, https://doi.org/10.1007/978-3-030-99125-8_2

2.2 Permutations: Refinement of Cycles' Coexistence

As was explained before, the Sh-theorem shows that periods of cycles *force* one another according to the Sh-order; here periods play the role of types of cycles. Notice that in the Sh-theorem periods are ordered linearly. However, in general, we do not require that the types of cycles always have the *linear forcing* property and allow for *partial forcing* too.

Let us formalize this observation. Consider the set of all cycles of $f \in C(I, I)$. One can categorize cycles into various types where by type one actually means a class of equivalence of cycles under some equivalence relation. In the setting of One-Dimensional *Combinatorial* Dynamics whether we include a cycle into a class of equivalence or not should depend upon the *combinatorics* of the cycle, but not, say, on smooth or metric properties of the map at points of the cycle. By combinatorics of the cycle we mean here the *cyclic permutation* naturally induced by the cycle in question. Also, given an equivalence relation $\sim$ we will call classes of $\sim$-equivalence $\sim$-*classes*.

Definition 2.1 If $A = \{x_1, \ldots, x_n\}$ is a cycle of an interval map f and $x_1 < x_2 < \cdots < x_n$, we say that a cyclic permutation $\psi_{A,f} : \{1, \ldots, n\} \to \{1, \ldots, n\}$ is *induced* by $f|_A$ if $f(x_i) = x_{\psi_{A,f}(i)}$ for each $i = 1, \ldots, n$. The family of all cycles on the real line that induce the same cyclic permutation σ is said to be a *(cyclic) pattern (defined by σ)* and is denoted by $\mathbf{P}_\sigma$.

We often omit the word "cyclic" and write "pattern" instead of "cyclic pattern". Patterns and cyclic permutations are in one-to-one correspondence; abusing the language we will use them interchangeably. Thus, $A \in \mathbf{P}_\sigma$ means that A is a cycle that induces a cyclic permutation σ; then we may say that A *is of (belongs to, has) pattern* $\mathbf{P}_\sigma$. Suppose that the space of all cyclic permutations $\mathscr{C}P$ is endowed with an equivalence relation $\sim$. Then, given a $\sim$-class $\mathscr{A}$, we denote by $\mathbf{P}_{\mathscr{A}}$ the family of all cycles on the real line that induce cyclic permutations from $\mathscr{A}$. If a cycle of a map f has pattern $\mathbf{A}$, say that f *exhibits* $\mathbf{A}$. If it does not cause ambiguity we write ψ_A instead of $\psi_{A,f}$.

Recall that a partial order "stronger" is *antisymmetric* if and only if the claim "a stronger than b" implies that the claim "b stronger than a" is false. Denote by CH(X) the convex hull of a set $X \subset \mathbb{R}$.

Definition 2.2 Consider the space $\mathscr{C}P$ of all cyclic permutations. Define a partial ordering $\gg$ on $\mathscr{C}P$ as follows: for cyclic permutations $\pi \neq \theta$ we write $\pi \gg \theta$ if and only if whenever a continuous interval self-mapping f has a cycle $E \in \mathbf{P}_\pi$, there must exist a cycle T of f with $T \in \mathbf{P}_\theta$. If $\pi \gg \theta$ then we say that π *forces* θ. The relation $\gg$ is called the *forcing relation* (or just the *forcing*) among cyclic permutations.

The proof of the next lemma uses a natural retraction.

Lemma 2.1 *If $\pi \gg \theta$ then for any continuous interval self-mapping f with a cycle $A \in \mathbf{P}_\pi$ there exists an f-cycle $B \in \mathbf{P}_\theta$ with $B \subset$ CH(A).*

Proof Define the canonical retraction R of the entire domain I of f onto $\mathrm{CH}(A)$ as the map that is identity on $\mathrm{CH}(A)$ and shrinks components of $I \setminus \mathrm{CH}(A)$ to their endpoints (shared by A). Consider the map $g = R \circ f : \mathrm{CH}(A) \to \mathrm{CH}(A)$. It is easy to see that cycles of g are in fact cycles of f. On the other hand, $\pi \gg \theta$ implies that g has a cycle $G \in \mathbf{P}_\theta$. Since by the above G is an f-cycle, we are done. $\square$

From now on in the situation when, for an interval map f, there is an f-cycle $A \in \mathbf{P}_\pi$ and $\pi \gg \theta$, we will always consider only f-cycles of pattern θ contained in $\mathrm{CH}(A)$.

Definition 2.3 (Canonical maps and cycles) For a cyclic permutation $\pi : \{1, \ldots, n\} \to \{1, \ldots, n\}$ let $f := f_\pi : K = [1, n] \to K = [1, n]$ be a piecewise-linear map defined as $f(i) = \pi(i)$ and linear on each $[i, i+1] = J_i$. We call f a *canonical piecewise-linear map generated by* π or just a π-*canonical map* and denote it by f_π. The f_π-cycle $A_\pi = \{1, \ldots, n\}$ belongs to $\mathbf{P}_\pi$ and is called the π-*canonical cycle*.

Definitions similar to Definition 2.3 can be given in a more general setting.

Definition 2.4 (Maps defined by a cycle) If $f : I \to I$ is an interval map and both endpoints of I belong to the same cycle T of f, then f is said to be T-*monotone* if it is monotone on each T-basic interval, and T-*linear* if it is linear on each T-basic interval. Thus, a π-canonical map is A_π-linear.

Let us now associate loops of A_π-basic intervals and cycles of f_π.

Definition 2.5 Let π be a cyclic permutation of period n. Suppose that $\pi \gg \theta$ where $\theta \neq \pi$ is a cyclic permutation of period m. By definition, f_π has a cycle $Q = \{q_1 < q_2 < \cdots < q_m\} \in \mathbf{P}_\theta$. Mark, for each i, an interval J_i containing q_i in its interior. Clearly, $f_\pi(J_i) \supset J_{i+1}, i = 1, \ldots, m-1$ and $f_\pi(J_m) \supset J_1$. The cycle Q is said to be *associated* to a loop of intervals $\widehat{J} = \{J_1 \to J_2 \to \cdots \to J_m \to J_1\}$.

Definition 2.5 is closely related to the concept of a cycle that follows a loop of intervals defined in Chap. 1. The difference is, in Chap. 1 it was shown that any loop L of P-basic intervals forces a periodic orbit that follows L. Here we move in the opposite direction, take a periodic orbit forced by a canonical cycle of a given pattern, and consider the loop associated to it.

In the loop of intervals $\widehat{J}$ from Definition 2.5, the images of the endpoints of each interval J_i are located to the left and to the right of the interior of the interval J_{i+1} and the images of the endpoints of the interval J_m are located to the left and to the right of the interior of the interval J_1. Notice also that there might exist several f_π-cycles of pattern $\mathbf{P}_\theta$, and, respectively, several loops of intervals like $\widehat{J}$.

Definition 2.6 (Essential loops) Let $\mathcal{O}$ be a periodic orbit of a continuous interval map f. A non-repetitive loop of $\mathcal{O}$-basic intervals not followed by a point of $\mathcal{O}$ is said to be *essential*.

In the setting of Definition 2.5, q_1 is a periodic point that follows $\widehat{J}$. By Lemma 1.4 from Chap. 1, an essential loop L of intervals defines a cycle that follows L. We claim that this is exactly the situation we are currently considering.

Lemma 2.2 *In the setting of Definition 2.5, the loop $\widehat{J}$ is essential.*

Proof Let us show that $\widehat{J}$ is non-repetitive. Indeed, otherwise for the least $s < m$ the loop of intervals $L = \{J_1 \to \cdots \to J_s\}$ is repeated k times in $\widehat{J}$ with $k > 1$ and $m = ks$. Choose a point $z \in J_1$ that follows L. Then $z \neq q_1$ because $s < m$. Assume that $z < q_1$. It follows that $f^m|_{[z,q_1]}$ is linear and such that $f^m(z) = z$, $f^m(q_1) = q_1$. Thus, $f^m|_{[z,q_1]}$ is the identity map. Since on each step the slope of f is at least 1, the only way it can happen is when $m = 2s$ and f^s isometrically "flips" $[z, q_1]$ to the left of z and then f^s isometrically "flips" it back onto $[z, q_1]$. This in turn implies that if $z \in [a, b]$ where $[a, b]$ is an A_π-basic interval then f^s isometrically "flips" $[a, b]$ so that $f^s(a) = b$ and $f^s(b) = a$. However then $\pi = \theta$, a contradiction.

Let us show that no point of A_π follows $\widehat{J}$. Suppose otherwise. Set $J_1 = [a, a + 1]$ and let the point a follow $\widehat{J}$. Then a and q_1 follow $\widehat{J}$, $f^m(a) = a$, $f^m(q_1) = q_1$, and since the map is linear on each A_π-basic interval, $f^m|_{[a,q_1]}$ is the identity. Hence, intervals $J_1, \ldots, J_m$ map onto each other and on each step the modulus of the slope is 1. Thus, $f^m(J_1) = J_1$, $f^m(a) = a$, and $f^m(a + 1) = a + 1$. Since $n \leq m$ is the period of A_π then $f^n(a) = a$, $f^n(a + 1) = a + 1$, and $f^n|_{[a,a+1]}$ is the identity. If $n < m$, then q_1 is of period $n < m$, a contradiction. Hence $m = n$, and $\pi = \theta$, a contradiction. $\square$

By Lemma 2.2 the fact that π forces θ implies that the π-canonical map f has a cycle B' of pattern θ that gives rise to an essential loop of intervals $\widehat{J}$ followed by B'. By Lemma 1.4 from Chap. 1, $\widehat{J}$ forces the existence of a periodic orbit B that follows $\widehat{J}$. Lemma 2.3 shows that $B = B'$.

Lemma 2.3 *Suppose that π is a cyclic permutation of period n and $\widehat{J} : J_1 \to J_2 \to \cdots \to J_m \to J_1$ is an essential loop of length m of A_π-basic intervals of the π-canonical map f_π. Then $f_\pi^m|_{J_1}$ is not the identity map, and if $B = \{q_1 < \cdots < q_m\}$ is a periodic orbit that follows $\widehat{J}$ then $B = B(\widehat{J})$ is unique.*

Proof Suppose otherwise. Let $J_1 = [a, a + 1]$. Then $f^m(a) = a$, $f^m(a + 1) = a + 1$, all intervals in $\widehat{J}$ map onto each other, and $m = kn \geq n$, $k \geq 1$. Since $\widehat{J}$ is non-repetitive, no A_π-basic interval shows in $\widehat{J}$ more than once. Since there are $n - 1$ intervals that are A_π-basic, then $m \leq n - 1$, a contradiction.

To prove the second statement, assume that there is a cycle $B' \neq B$, $B' = \{q_1' < q_2' < \cdots < q_m'\}$ of f_π that follows $\widehat{J}$. Assume for definiteness that $q_1' < q_1$. Since both q_1 and q_1' follow the same loop of intervals, the map $f^m|_{[q_1',q_1]}$ is the identity map. As before this implies that $f(J_i) = J_{i+1}$ for each $i = 1, \ldots, m - 1$ while $f(J_m) = J_1$. Since $f^m|_{[q_1',q_1]}$ is the identity map, $f^m|_{J_1}$ is the identity map, a contradiction with the above. $\square$

We have shown that if a cyclic permutation π forces a cyclic permutation θ then there exists an essential loop $\widehat{J}$ of A_π-basic intervals under the π-canonical map f_π such that the unique (by Lemma 2.3) f_π-cycle $B(\widehat{J})$ that follows $\widehat{J}$ belongs to the pattern $\mathbf{P}_\theta$. Let us now prove the opposite: Given any essential loop $\widehat{J}$ of A_π-basic intervals under the π-canonical map f_π, the permutation θ induced by $B(\widehat{J})$ is forced by π.

Theorem 2.1 *If $\widehat{J} : J_1 \to J_2 \to \cdots \to J_m \to J_1$ is an essential loop of A_π-basic intervals of the π-canonical map f_π and θ is the cyclic permutation induced by the cycle $B(\widehat{J})$ then π forces θ.*

We use the following notation: given a cyclic permutation π, an interval map g, a g-cycle P of pattern π, and an A_π-basic interval J, denote by J^P the P-basic interval corresponding to J, and use the same agreement for loops of intervals. Thus, the leftmost A_π-basic interval J is associated with the leftmost P-basic interval J^P, and so on. Observe that if $\widehat{J} : J_1 \to \cdots \to J_m \to J_1$ is an (essential) loop of A_π-basic intervals then $\widehat{J}^P : J_1^P \to \ldots J_m^P \to J_1^P$ is an (essential) loop of P-basic intervals. In what follows, we prove a sequence of lemmas in which we use the notation from the statement of Theorem 2.1; in particular, we consider an essential loop of A_π-basic intervals $\widehat{J} : J_1 \to \ldots J_m \to J_1$.

Definition 2.7 If a loop of A_π-basic intervals $\widehat{J} : J_1 \to \cdots \to J_m$ is such that $f_\pi(J_i) = J_{i+1}, i = 1, \ldots, m-1$ and $f_\pi(J_m) = J_1$ then $\widehat{J}$ is called *non-expanding*; the loop $\widehat{J}^P$ is called *non-expanding* too. Otherwise there is $1 \leq i \leq m-1$ such that $f_\pi(J_i)$ contains J_{i+1} and at least one more A_π-basic interval adjacent to J_{i+1}. Then $\widehat{J}$ is called *expanding*; the loop $\widehat{J}^P$ is called *expanding* too.

We begin by considering the non-expanding case.

Lemma 2.4 *Theorem 2.1 holds for non-expanding loops.*

Proof By Lemma 2.3, $f_\pi^m|_{J_1}$ is not the identity map; hence, $f_\pi^m|_{J_1}$ is an isometric flip of J_1 onto itself. Clearly, all intervals $J_1, \ldots, J_{m-1}$ are distinct as otherwise our assumptions on $\widehat{J}$ would imply that $\widehat{J}$ is repetitive, a contradiction with $\widehat{J}$ being essential. If a continuous interval map F has a cycle $X \in \mathbf{P}_\pi$ then the loop of intervals $\widehat{J}^X = \{J_1^X \to \cdots \to J_m^X \to J_1^X\}$ is essential. By Lemma 1.4 from Chap. 1, there is an F-cycle that follows $\widehat{J}^X$. Since all intervals $J_1, \ldots, J_{m-1}$ are distinct then this cycle induces θ as desired. $\qquad\square$

We will need the next definition where notions related to piecewise-monotone maps are introduced.

Definition 2.8 Consider an interval map $h : I \to I$ with finitely many points $t_1, \ldots, t_k$ in the interior of I such that h restricted on any component of $I \setminus \{t_1, \ldots, t_k\}$ is strictly monotone and monotonicity on any two adjacent components is distinct. Then we call the points $t_1, \ldots, t_k$ *turning points* of h and the components of $I \setminus \{t_1, \ldots, t_k\}$ the *laps* of h. All such maps h will be called *strictly piecewise monotone*.

Evidently, if h is strictly piecewise monotone then h^s is also strictly piecewise monotone, its turning points are all preimages of turning points of h under $h, \ldots, h^{s-1}$, and the laps are, as always, closures of components of I with removed turning points of h^s.

Definition 2.9 Say that an interval K *follows* a loop of intervals $\widetilde{J} : J_1' \to \cdots \to J_t' \to J_1'$ defined for a map g if $K \subset J_1'$, $g(K) \subset J_2'$, $\ldots$, $g^{t-1}(K) \subset J_t'$, $K \subset g^t(K) \subset J_1'$.

Let us now consider expanding loops of intervals.

Lemma 2.5 *Suppose that $\widehat{J}$ is expanding. Then there is a number N such that the lap E of f_π^{Nm} that follows $\widehat{J}$ is such that $E, f_\pi(E), \ldots, f_\pi^{m-1}(E)$ are pairwise disjoint while $f_\pi^m(E) \supset E$. In particular, the spatial order among the intervals $E, f_\pi(E), \ldots, f_\pi^{m-1}(E)$ mimics the spatial order of points of the cycle $B(\widehat{J})$.*

Proof Consider the lap T of f_π^m that follows $\widehat{J}$ such that $f_\pi^m(T) = J_1$ (clearly, T can be found if we pull J_1 back along $\widehat{J}$). By the assumptions, T linearly expands onto J_1 under f_π^m, and T is a subset of the interior of J_1 (otherwise an endpoint of J_1 follows $\widehat{J}$ contradicting $\widehat{J}$ being essential). Iterating $f_\pi^m|_T$ and pulling T back into itself we obtain a nested sequence of laps of f_π^m, f_π^{2m}, etc. shrinking to a periodic point b of period m. The periodic orbit B of b is that unique periodic orbit of $f_p i$ that follows $\widehat{J}$ and whose uniqueness was established in Lemma 2.3. If N is large, the lap E of f^{Nm} defined above is such that $E, f(E), \ldots, f^{m-1}(E)$ are pairwise disjoint while $f^m(E) \supset E$.   □

The idea of the proof of Theorem 2.1 is to construct intervals similar to E for an arbitrary continuous map g that exhibits $\mathbf{P}_\pi$ and prove that this implies the existence of a cycle of g that belongs to the pattern $\mathbf{P}_\theta$. The proof of Lemma 2.6 is left to the reader.

Lemma 2.6 *Suppose that $f : \mathbb{R} \to \mathbb{R}$ is such that for some points $a < b$ we have that either (1) $f(a) = t_0 < t_1 < t_2 < \cdots < t_n = f(b)$ or (2) $f(a) = t_0 > t_1 > \cdots > t_n = f(b)$. Then there are points $s_0 = a < s_1 < s_2 < \cdots < s_{n-1} < s_n = b$ such that $f(s_i) = t_i$ for each $i = 0, \ldots, n$.*

Lemma 2.6 allows one to deal with continuous interval maps in almost the same fashion one could deal with piecewise monotone or even piecewise-linear maps. In particular, Lemma 2.6 implies Lemma 2.7. Let π' be a cyclic permutation. Consider the set of all turnings points $T = \{t_1, \ldots, t_N\}$ of $f_{\pi'}^s$ (these include turning points of $f_{\pi'}$); clearly, T is an $f_{\pi'}$-invariant set such that $f_{\pi'}^{s-1}(T) = A_{\pi'}$. Lemma 2.7 follows from Lemma 2.6 by induction on s.

Lemma 2.7 *Let π' be a cyclic permutation. Consider the set T of all turning points $t_1, \ldots, t_N$ of $f_{\pi'}^s$. If $F : I \to I$ is an interval map with a cycle $P \in \mathbf{P}_{\pi'}$, then there is an F-invariant set T^P of points in I that contains P and is such that $F|_{T^P}$ is monotonically increasingly conjugate to $f_{\pi'} : T \to T$.*

We are ready to prove Theorem 2.1.

Proof of Theorem 2.1. We use the notation from the statement of the theorem. By Lemma 2.4 it holds for non-expanding loops of A_π-basic intervals. Assume now that $\widehat{J}$ is expanding. Then by Lemma 2.5 we can find a number N such that a lap E of f_π^{Nm} that follows $\widehat{J}$ is such that $E, f_\pi(E), \ldots, f_\pi^{m-1}(E)$ are pairwise disjoint while $f_\pi^m(E) \supset E$. In particular, the spatial order among the intervals $E, f_\pi(E), \ldots, f_\pi^{m-1}(E)$ mimics the spatial order among points on which the permutation θ acts.

Now, suppose that an interval map $F : I \to I$ has a cycle $P \in \mathbf{P}_\pi$. Let T be the set of all turning points of f_π^{Nm}. Then by Lemma 2.7 there exists an F-invariant set T^P of points in I and a monotonically increasing map $\psi : T \to T^P$ that conjugates $f_\pi|_T$ and $F|_{T^P}$. Choose an interval E_1^P, a component of $I \setminus T^P$, whose endpoints are ψ-images of the endpoints of E. Moreover, construct intervals $E_2^P, \ldots, E_m^P$ analogously. Then the endpoints of each E_i^P are the F^{i-1}-images of the endpoints of E_1^P. Evidently, intervals $E_1^P, \ldots, E_m^P$ are almost disjoint; moreover, $E_1^P \to \ldots E_m^P \to E_1^P$ form a loop of intervals under F which implies (by Lemma 1.4 from Chap. 1) the existence of an F-periodic point $y \in E_1^P$ of period m that follows this loop of intervals. Evidently, the orbit of y induces the cyclic permutation θ as desired. $\qquad\qquad\square$

Theorem 2.1 shows that essential loops of A_π-basic intervals correspond to cyclic permutations forced by π. Since these loops of intervals can be mimicked for any T-linear map where T is a cycle inducing π, Theorem 2.1 can be restated as follows (the details can be found, e.g., in [2]).

Theorem 2.1a *The cyclic permutations forced by a cyclic permutation π are exactly the cyclic permutations exhibited by any T-linear map where T induces π.*

Let us show that the relation $\gg$ is antisymmetric.

Lemma 2.8 ([4]) *The relation $\gg$ is antisymmetric.*

Proof Let $\pi \gg \theta$ and $\theta \gg \pi$. Assume that π is of order n and consider the map $f_\pi : [1, n] \to [1, n]$. There must exist an f_π-cycle $D_1 \in \mathbf{P}_\theta$. By Lemma 2.1 we can then find a cycle $A_1 \subset \mathrm{CH}(P)$ with $A_1 \in \mathbf{P}_\pi$, etc. This yields a sequence of cycles D_i, A_i of f_π whose convex hulls are monotonically decreasing and whose patterns alternate between $\mathbf{P}_\theta$ and $\mathbf{P}_\pi$, respectively. We may assume that the convex hulls of these cycles form a nested sequence of closed intervals converging to a closed interval $I = [u, v] \subset (1, n)$. Thus, u and v are the limits of sequences of leftmost and rightmost points of cycles D_i. This implies that the orbit B of u is, say, m-periodic and contains v. Clearly, $1 \notin B$ and $n \notin B$. Hence, B is disjoint from A_π and the map f_π^m is linear on a small interval U containing u.

Now, if $f_\pi^m|_U$ is expanding then, clearly, all points close to u are either non-periodic or periodic with periods growing to infinity. This contradicts the assumptions about cycles P_i, A_i (whose periods are bounded). Hence $f_\pi^m|_U$ is with slope 1 or -1. In either case all points close to u are periodic with the periodic orbits from the same pattern (equivalently, they all induce the same cyclic permutation). However by construction arbitrarily close to u there exist periodic points whose orbits either belong to $\mathbf{P}_\pi$ or to $\mathbf{P}_\theta$ where $\pi \neq \theta$, a contradiction. $\qquad\square$

The antisymmetry of $\gg$ yields the following "variational" concept.

Definition 2.10 If a cyclic permutation π of period n does not force any other cyclic permutations of the same period we will call π *forcing-minimal*.

We will rely upon Theorem 2.1 in the next definition.

Definition 2.11 Let the space of all cyclic permutations $\mathscr{C}P$ be endowed with equivalence relation $\sim$. Define a partial ordering $\gg_\sim$ as follows: for every two $\sim$-classes $\mathscr{E}$ and $\mathscr{T}$, $\mathscr{E} \gg_\sim \mathscr{T}$ if and only if for every permutation $\pi \in \mathscr{E}$ there exists a permutation $\theta \in \mathscr{T}$ such that π forces θ. If $\gg_\sim$ is antisymmetric, we say that $\gg_\sim$ is of *forcing type*.

It follows that $\gg$ is $\gg_=$ where $=$ is, of course, the degenerate equivalence relation (where each cyclic permutation is equivalent only to itself). By Lemma 2.8 $\gg$ is of forcing type. Also, it is easy to see that $\gg_\sim$ is transitive. Moreover, if a $\sim$-class $\mathscr{A}$ contains a trivial cyclic permutation of period 1 then $\mathscr{A} \gg_\sim \mathscr{B}$ is impossible for any $\sim$-class $\mathscr{B}$.

Lemma 2.9 *If all $\sim$-classes are finite then the relation $\gg_\sim$ is antisymmetric.*

Proof This follows from the antisymmetry of $\gg$ (Lemma 2.8); the details are left to the reader.   □

The Sh-theorem uses the equivalence relation defined by the period, i.e., by the order of permutations; by Lemma 2.9 the corresponding order (i.e., the Sh-order) is of forcing type. Other forcing type relations among cyclic patterns may follow with various choices of equivalence relations on cyclic permutations.

As an application, let us deduce an important theorem of L. Block from the results of Chap. 1 using the properties of forcing relation among patterns established in this section.

Theorem 2.2 ([8, 14]) *If $f \in C(I, I)$ has a cycle of period n, then f has also a simplest cycle of period n.*

Proof Take any pattern of period k. It may force other patterns of period k which in turn may force other patterns of period k, etc. Hence any pattern of period k forces at least one Sh-weakest pattern of period k. Denote this pattern by θ, take a cycle P of pattern θ, and consider the P-linear map f. By the choice of P and by Theorem 1a, P is a unique cycle of f of period k. We claim that k is the Sh-strongest period of cycles of f. Indeed, otherwise f has a cycle Q of some period m such that $m \succ k$. Denote the pattern of Q by π. By the SH-theorem the pattern π forces a pattern $\tilde{\theta}$ of period k. By Lemma 2.1 this implies that there exists an f-cycle T of pattern $\tilde{\theta}$ contained in the convex hull CH(Q) of Q. Clearly, $T \cap P = \emptyset$. Thus, f has another cycle of period k distinct from P, a contradiction.

Thus, indeed, k is the Sh-strongest period among all periods of cycles of f. By the results of Subsect. 1.2.5 of Chap. 1, the pattern θ is simplest. In other words, any pattern of period k forces a simplest pattern of period k. Hence, any map $f \in C(I, I)$ that has a cycle of period n must also have a simplest cycle of period n.   □

The relation $\gg$ among cyclic patterns was first introduced in [4] and [3] and then developed in a number of papers and books, including [32] and [2]. It follows from [4] that forcing is a partial order among patterns. However, this partial order has features quite opposite to those of periods: it is too fine. As a result, a concise and explicit description of the set of patterns forced by a given pattern is not known. This motivates a quest for a different, "middle-of-the-road" notion which would be not as refined as patterns but not as crude as periods of cycles. A possible choice of such notion is considered in the next section.

2.3 Rotation Theory

The Sh-theorem uses a specific order among the periods of cycles of an interval map. One can think of periods having different strengths, so that stronger periods *force* weaker periods (to be among periods of cycles of an interval map). Simply put, periods force periods. The description of possible sets of periods of cycles of continuous interval maps ensues. However, there are finer than periods but still numerical ways to describe interval cycles (i.e., by a fixed number of integers—say, two—however long the cycle is). The concepts of rotation pair/number [16, 19] fit into this description.

Here it turns out to be very useful to rely upon the concept of the number of rotations per period which we will now introduce. The movement on $\mathbb{R}$ is to the right or left. Let $A = \{a_1 < a_2 < \cdots < a_k\}$ be points of a cycle and let a map f act on these points according to a cyclic permutation $i_1 = 1 \to i_2 \to i_3 \to \ldots \to i_k \to 1$: First $a_{i_1} = a_1$ moves to the right to a_{i_2}, then a_{i_2} maps in some direction to a_{i_3}, etc. Each time the direction of the point's movement changes, it can be visualized as the turn (rotation) of the point by 180 degrees (or, equivalently, by $\frac{1}{2}$ of the full angle) *in the positive direction*. Evidently, two 180-degree rotations add up to one (full) rotation.

Similarly, one can consider rotation of a vector from x to $f(x)$ where $x \in A$. Then the initial vector is from a_{i_1} to a_{i_2}, then we get the vector from a_{i_2} to a_{i_3}, etc. Each time the direction of the vector changes, the vector rotates by 180 degrees (or by $\frac{1}{2}$ of the full angle) in the positive direction. After k steps the point comes back to itself (equivalently, the vector comes back onto itself). The cumulative rotation p is just the number of rotations per period corresponding to $2p$ changes of the direction of the movement of a point (or $2p$ rotations of the vector).

One can compute the cumulative rotation as follows. Associate to each number $j, 1 \le j \le k$ the sign $+$ if $f(j) > j$ and the sign—otherwise. Then associate to the permutation in question a finite string of pluses and minuses, beginning with the sign associated to a_1 and then following (temporally) the orbit. In this string of signs, we can ignore segments of either pluses or minuses and replace each such string by one appropriate sign. The remaining string will consist of alternating pluses and minuses, and its length will be $2p$.

Define the *rotation pair* of a cycle as (p, q), where q is the period of the cycle and p is the number of rotations per period of this cycle. The number $p/q \le 1/2$ is called the *rotation number* of the cycle. Let us introduce the following partial ordering among all pairs of integers (p, q). Write $(p, q) \succ (r, s)$ if $p/q < r/s \le 1/2$ or $p/q = r/s = m/n$ with m and n coprime and $p/m \succ r/m$ (notice that $p/m, r/m \in \mathbb{N}$).

Theorem 2.3 ([16, 19]) *If* $f : [0, 1] \to [0, 1]$ *is continuous and has a cycle of rotation pair* (p, q) *then* f *has cycles of any rotation pair* (r, s) *such that* $(p, q) \gg (r, s)$.

This theorem can be understood in the sense of *forcing* among rotation pairs of interval cycles: the fact that (p, q) is the rotation pair of a cycle of a map f *forces* the presence of other cycles of f with every rotation pair (r, s) such that $(p, q) \gg (r, s)$. Evidently, Theorem 2.3 is modeled after the Sh-theorem. Moreover, Theorem 2.3 implies a full description of the sets of rotation pairs for continuous maps; as in the Sh-theorem, all theoretically possible sets really occur.

The paper [19] is inspired by [16] where similar but less productive notions were introduced. Compared to [16], the article [19] gains in simplicity of both results and proofs, in part because it deals with rotation pairs and numbers. Both [16] and [19] are inspired by [30] where sets of similar pairs of numbers (reflecting the number of rotation per period and period itself) are described for circle maps of degree one. Notice that analogous phenomena were also discovered in [23].

One example: suppose that we know that an interval map f has a cycle of period, say, 11; then according to the Sh-theorem we can only guarantee that it has cycles of periods 13, 15, etc. However, we cannot easily say in what cases the existence of a cycle of period 11 forces the existence of cycles of periods, say, 3, or 5, or 7, or 9; to this end we need to study the permutation induced by the cycle so that from purely numerical the problem becomes combinatorial.

Assume now, that there exists an f-cycle of rotation pair, say, $(2, 11)$. Then not only can we guarantee that f has cycles of periods 9, 7, 5, and 3 but also that some of these cycles have rotation pairs $(2, 9)$, $(3, 9)$, $(4, 9)$, $(2, 7)$, $(3, 7)$, $(1, 5)$, $(2, 5)$, and $(1, 3)$. Thus, since we are now using more informative input we are getting a slightly richer output. Also, Theorem 2.3 and the definition of the order $\gg$ are easy to follow as both are related to the order of rotation numbers (all of whom must be less than or equal to $1/2$) with respect to their distance to $1/2$; this order is rather transparent and easy to grasp.

Forcing for rotation pairs is given by the relation $\gg$. Thus, using rotation pairs we get a situation that is not much more complicated than that for periods, but digs deeper into the structure of cycles. Notice also, that if Q is an f-cycle of rotation pair (m, n), and we conjugate the map f restricted on Q to another map g on a g-cycle Q' by a homeomorphism that reverses the orientation then the rotation pair and rotation number of Q' do not change. This shows advantages of Theorem 2.3.

On the other hand, the proof of Theorem 2.3 uses the Sh-theorem in a substantial way (together with a few new arguments and concepts). We conclude that Theorem 2.3 is an interesting and natural development of the ideas and tools involved in the proof of the Sh-theorem.

While rotation pairs are a finer characteristic of permutations than periods alone, they are not as precise as permutations themselves. The forcing relation among permutations with the same rotation pairs is non-trivial. For example, continuous interval maps may have cycles of period 4 and the same rotation pair $(1, 4)$ which exhibit three distinct cyclic permutations, namely, $1 \to 2 \to 3 \to 4 \to 1$, $1 \to 2 \to 4 \to 3 \to 1$, and $1 \to 3 \to 4 \to 2 \to 1$. It is not hard to see that none of the patterns associated with the above-listed permutations forces another one. Moreover, the first pattern is unimodal, the second one is bimodal of the "increasing-decreasing-increasing" (N-)type, and the third one is bimodal of the "decreasing-increasing-decreasing" (U-)type. However, already for the rotation pair $(2, 5)$ there are two distinct cyclic permutations, $1 \to 3 \to 4 \to 2 \to 5 \to 1$ and $1 \to 2 \to 4 \to 3 \to 5$, and it is not hard to verify that the pattern associated with the second permutation forces that associated with the first one.

Thus, given an rotation pair (p, q) there is a variety of cycles with this rotation pair. These cycles vary according to their number of intervals of monotonicity and other parameters. There may exist non-trivial forcing relation among some patterns of rotation pair (p, q). Still, the information provided by Theorem 2.3 helps us describe variety of cycles of interval maps slightly beyond the information provided by the Sh-theorem. This is because the rotation pair of a cycle is a finer characteristic of a cycle compared to its period.

Let us now sketch the proof of Theorem 2.3.

Lemma 2.10 *If f has a cycle with points $x < y$ such that $f(x) < x$ and $f(y) > y$ then all rotation pairs are possible.*

Call a cycle *divergent* if it has points $x < y$ such that $f(x) < x$ and $f(y) > y$. A cycle that is not divergent is called *convergent*. Lemma 2.10 is based upon the analysis of divergent cycles discovered in the original proof of the Sh-theorem.

Definition 2.12 (L-scheme) Suppose that there are points a, y, z such that $f(a) = a$ and $f(z) \leq a < y < z \leq f(y)$ (alternatively, there are points z', y', a' such that $f(a') = a'$ and $f(y') \leq z' < y' < a' \leq f(z')$). Then we say that f has an L-scheme.

Lemma 2.11 *Suppose that, up to orientation, there exists a fixed point a and a point x such that $a < x < f(x)$ and, for some $n > 1$, we have $f^n(x) \leq a$. Then f has an L-scheme. In particular, this holds if f has a divergent periodic orbit P. Moreover, if a map has an L-scheme, then it has cycles of all periods.*

Proof Let us choose the closest to x from the left fixed point a' of f. Clearly, $a \leq a' < x$ and all points between a' and x map to the right under f. Choose the least m such that $f^m(x) \leq a'$; then $1 < m \leq n$ and, in addition to that, $a' < f^{m-1}(x)$. Now, choose the first time k when x is mapped to the right of $f^{m-1}(x)$ (non-strictly); then $a' < f^{k-1}(x) < f^{m-1}(x) \leq f^k(x)$. If we set $f^{k-1}(x) = y$ and $f^{m-1}(x) = z$, we see that $f(z) \leq a' < y < z \leq f(y)$ as required in Definition 2.12).

Let us show that if f has a divergent periodic orbit P then the above conditions are satisfied. Indeed, we may assume that $u < v$ are adjacent points of P such that $f(u) < u$ and $f(v) > v$. Then there exists a fixed point $a \in (u, v)$. Let us follow P from v on; then there exists m such that $f^m(v) = u < a$. Thus, the above conditions hold for v and a.

Finally, let us show that the presence of an L-scheme implies the existence of cycles of all periods. Indeed, assume that $f(z) \leq a = f(a) < y < z \leq f(y)$. We may assume that for any $x \in (a, y]$ we have $x < f(x)$. Then we can choose the least $z_{-1} \in (a, y]$ with $f(z_{-1}) = z$, then the least $z_{-2} \in (a, z_{-1})$ with $f(z_{-2}) = z_{-1}$, and so on inductively. Let us show that then there is a periodic point of period k in (z_{-k}, z_{-k+1}). Indeed, by construction $f^k(z_{-k}) = z = z_0$, $f^k(z_{-k+1}) = f(z_0) \leq a$, and for any point $x \in (z_{-k}, z_{-k+1})$ we have that $x < f(x) < z_{-k+2}$, $x < f(x) < f^2(x) < z_{-k+3}, \ldots, x < \cdots < f^{k-1}(x) < z_0$. Hence, the period of x is exactly k as desired. $\square$

This analysis shows that divergent cycles force the existence of points a, y, z such that $f(a) = a$ and $f(z) \leq a < y < z \leq f(y)$ (alternatively, there are points z', y', a' such that $f(a') = a'$ and $f(y') \leq z' < y' < a' \leq f(z')$). This implies that intervals $[a, y]$ and $[y, z]$ have images covering their union which, in turn, implies the claim made in Lemma 2.10.

Forcing of cycles of interval maps can be detected if, given a cycle P of a map f, one considers not f but the P-*linear* map g which coincides with f on P, is otherwise linear, and is defined on the convex hull T of P (see, e.g., [4], or [2], or Sect. 2.2). So, take a map f with a cycle P. If P is divergent then by Lemma 2.10, Theorem 2.3 holds. Let us now assume that P is convergent and, by [4] and [2], that f is P-linear and defined on the convex hull T of P. Then f has a unique fixed point, i.e., $f \in \mathcal{U}_T$, where $\mathcal{U}_T$ is the family of all maps from $C(T, T)$ having a unique fixed point (we will always denote this fixed point by a). Observe that every cycle of $f \in \mathcal{U}_T$ is convergent, and if $f \in \mathcal{U}_T$ then the first element in the rotation pair of an f-cycle Q is the number of points $x \in Q$ such that $f(x) < a < x$.

For any map $f \in \mathcal{U}_T$ we use loops of intervals of a special type. Call an interval with one of the endpoints equal to a *admissible*. Let I be admissible; if it is located to the left of a, set $\phi(I) = 0$, otherwise set $\phi(I) = 1$. A loop of admissible intervals is called *admissible*. An admissible loop $\widetilde{J}$ has *rotation pair* $(p, q) = \text{orp}(\widetilde{J})$ if it has length q and in it there are p blocks $J \to K$ with J to the right of a and K to the left of a. Then p/q is called the *rotation number* of $\widetilde{J}$ and is denoted $\rho(\widetilde{J})$. Loops that have a common interval can be concatenated.

Lemma 2.12 *Let $f \in \mathcal{U}_T$ and $\bar{\alpha} = \{I_0, \ldots, I_{k-1}\}$ be an admissible loop. If k is even and $\phi(I_j) = 0$ for even j and $\phi(I_j) = 1$ for odd j, then f has a point x of period 2. Otherwise there is a point $x \in I_0$ with $x \neq a$, $f^j(x) \in I_j (0 \leq j \leq k-1)$, $f^k(x) = x$ and so $\rho(x) = \rho(\bar{\alpha})$. Moreover, if the sequence of numbers $\{\phi(I_0), \ldots, \phi(I_{k-1})\}$ is non-repetitive then $\text{orp}(x) = \text{orp}(\bar{\alpha})$.*

Proof Without loss of generality we may assume that $T = [0, 1]$. Suppose that the first case from the lemma holds but f has no points of period 2. By the Sh-theorem then all points of $[0, 1]$ except for a are non-periodic. Hence, $f^i(0) > 0$ for any $i > 0$, which implies that $f^i(x) > x$ for any $x \in [0, a)$ (otherwise $f^i(0) > 0$, $f^i(x) < x$ and there must exist a point $y \in (0, x)$ with $f^i(y) = y$, a contradiction). This makes $f^k(I_0) \supset I_0$ impossible. If now the second case holds, then some I_j and I_{j+1} lie both to the same side of a. As $f(I_j) \subset I_j$ is impossible (it would rule out $f^k(I_j) \supset I_j$), it follows that we can find an interval $E \subset I_j$ with $f(E) = I_{j+1}$ and $a \notin E$. Replace I_j by E in $\bar{\alpha}$ to construct a new loop $\bar{\beta}$ and find, using the tools from Chap. 1, a periodic point $x \in E$ that follows $\bar{\beta}$. Then, clearly, $x \neq a$ is not fixed, and it is easy to see that x has the desired properties. $\square$

Assume that $f \in \mathscr{U}_T = \mathscr{U}$ (e.g., f may be P-linear for some convergent cycle P with T being the convex hull of P). If we denote by J_z the interval with endpoints a and z then for $x \in P$, $J_x \to J_{f(x)} \to \cdots \to J_{f^{n-1}(x)} \to J_x$ is an admissible loop called the *fundamental admissible loop of P*. Evidently, if we concatenate loops of rotation pairs (p, q) and (r, s) then we get a loop of rotation pair $(p + r, q + s)$.

Lemma 2.13 *If $f \in \mathscr{U}$ has a cycle of rotation number α and $\beta \in [\alpha, 1/2]$ is rational then f has a cycle of rotation number β.*

Proof Let P be an f-cycle of rotation number α and period n, let $\beta \in [\alpha, 1/2]$ be rational. Then there are non-negative integers r, s such that $(rn\alpha + s)/(rn + 2s) = \beta$. Let x and y be the points of P closest to a from the left and right, respectively. Since x is mapped to the right and y to the left, $J_x \to J_y \to J_x$ is an admissible loop. The concatenation of s copies of this loop with r copies of the fundamental admissible loop of P (J_x appears in both loops!) is a loop of rotation number β. By Lemma 2.12 f has a cycle of rotation number β. $\square$

The next lemma is central in the proof of Theorem 2.3.

Lemma 2.14 *Let p, q be coprime. If $f \in \mathscr{U}$ has a cycle of rotation number p/q, then f has a cycle of rotation pair (p, q).*

Proof Let $f \in \mathscr{U}$ and let P be a cycle of the smallest period among the f-cycles of rotation number p/q. The rotation pair of P is (mp, mq) for some $m \geq 1$. Suppose that $m > 1$. Consider the fundamental loop of P and compute the first element of the rotation pair of P by adding $1/2$ for every arrow that starts and ends on opposite sides of a. Look at such sums for q consecutive arrows of the loop. If we move with our block by one arrow along the loop then this sum changes by at most $1/2$ (as we move our block by one arrow, we lose its initial arrow and add a new terminal arrow, hence we subtract at most $1/2$ and then add at most $1/2$ which changes the sum of $1/2$'s involved in the q-tuple of arrows by at most $1/2$).

Since the average sum over such blocks is p, then there is a block over which that sum is exactly p. Indeed, otherwise there are sums that are less than p, and there are sums that are greater than p. However, each sum is an integer multiple of $1/2$, and on each step, the sum changes by at most $1/2$. Hence, as we step by step move from a sum that is less than p to a sum that is greater than p, we will inevitably find a sum that is equal to p. One can view this as a discrete version of the Intermediate Value Theorem.

Take a block with the sum p. It starts with, say, J_x and ends with J_y. Computing its sum, we add $1/2$ each time we move across a. Since this sum is an integer, x and y must lie on the same side of a. Therefore either $J_x \subset J_y$ or $J_y \subset J_x$. Hence, either our block forms a loop of rotation pair (p, q) or its complement to the fundamental loop of P forms a loop of rotation pair $((m - 1)p, (m - 1)q)$. This contradicts minimality of period of P. Thus, $m = 1$, so the rotation pair of P is (p, q). $\square$

These lemmas prove Theorem 2.3 for coprime rotation pairs. Since we only consider convergent cycles, we may assume that a map f belongs to $\mathscr{U}$. If f has a cycle of rotation pair (mp, mq) where P and q are coprime, then by Lemma 2.14 f has a cycle of coprime rotation pair (p, q). Moreover, by Lemma 2.13 for any coprime pair (m, n) with $p/q < m/n \leq 1/2$ there exists an f-cycle of rotation pair (sm, sn). However, we can again use Lemma 2.13 that implies that f has a cycle of rotation pair (m, n).

To complete the sketch of the proof of Theorem 2.3 we need a few definitions. Say that a cycle Q has a *block structure (over a cycle P)* if P is of period $n > 1$, Q is of period kn with $k > 1$ (cf. [32]), and for the Q-linear map f there are pairwise disjoint intervals $K_1, K_2, \ldots, K_n$ such that $f(K_i) \subset K_{i+1}$ for $i < n$, $f(K_n) \subset K_1$, each interval K_i contains a unique point p_i of P, and $g(p_i) = p_{i+1}, g(p_n) = p_1$. Notice that since K_i are pairwise disjoint, there are k points of Q in each K_i. Sets $Q \cap K_i$ are called *blocks (of Q)*. If no cycle P is given while Q has all the listed properties, we simply say that Q has *block structure*. If each block consists of two points we say that Q is a *doubling*.

A crucial tool in what follows is a nice and powerful result due to Misiurewicz and Nitecki, namely, Theorem 9.12 [32]. We prefer its version given in Theorem 2.4 in [20] and restated in a convenient for us form. Suppose that a pattern π has a block structure over a pattern θ. Moreover, suppose that on each block except for one the map is monotone, on exactly one block the map is unimodal, and first return map on a block induces a certain pattern ξ. Then we say that π is an extension of θ by a pattern ξ. Similar definitions can be given for cycles.

Theorem 2.4 (Theorem 2.4 [20]) *Suppose that an A-linear map f has a cycle B, A has no block structure over B, and B is not a doubling. Suppose that the rotation pair of B is (p, q). Then f has cycles that are extensions of B by any unimodal pattern. In particular, f has cycles of rotation pairs (ip, iq).*

Consider a map $f \in \mathcal{U}$. Suppose that P is a cycle of f of rotation pair (p, q). Let (m, n) be a coprime pair with $p/q < m/n < 1/2$ (in other words, let $(p, q) > (m, n)$). Consider the P-linear map g. Then by the above arguments g has a cycle Q of rotation pair (m, n). Since $p/q < m/n < 1/2$ and (m, n) is coprime, then it is easy to see that P has no block structure over Q, and Q itself is not a doubling. Then the rest of Theorem 2.3 in that case follows from Theorem 2.4. The remaining cases of Theorem 2.3 are proven similarly, but are slightly more technical and are omitted here.

The rotation theory of interval maps is still the work in progress, with a number of interesting open questions. Here is one of them. Given an interval map f, consider the closure $I_{rot}(f)$ of the set if all rotation numbers of its cycles. Then, by Lemma 2.13, $I_{rot}(f)$ is a closed interval $[\rho_f, \frac{1}{2}]$. Let us call ρ_f the *rotational capacity* of f. The less it is, the greater the set of rotation pairs of cycles of f. Thus, the rotational capacity can be viewed as a measure of richness of its family of cycles. A natural question, parallel to a similar question for the topological entropy, is whether the iso-rotational capacity sets of interval maps are connected (all continuous maps, or maps of specific type or modality). In other words, given a space of interval maps $\mathscr{A}$ and a number s, is the family of maps $f \in \mathscr{A}$ with $\rho_f = s$ connected?

An interesting application of the rotation theory can be found in recent papers [21, 22] in which cyclic patterns on the intervals were considered from a structural point of view. Let us briefly describe results of these papers.

Definition 2.13 (Block structure) Let π be a cyclic permutation of the set $X = \{1, \ldots, n\}$. Suppose that for some $k > 1$ and $m > 1$ we have $n = km$ and the permutation π maps set $Y_1 = \{1, \ldots, m\}$, $Y_2 = \{m + 1, \ldots, 2m\}$, $\ldots$, $Y_k = \{n - m + 1, \ldots, n\}$ to one another. Then sets $Y_1, \ldots, Y_k$ are called *blocks* and the permutation π is said to have *block structure*; call k the *period* of this block structure. Similar terminology is used for patterns and cycles. Otherwise, a permutation (a pattern, a cycle) is said to have *no block structure*.

Define the following order among all natural numbers:

$$4 \gg 6 \gg 3 \gg \cdots \gg 4n \gg 4n + 2 \gg 2n + 1 \gg 4n + 4 \gg \cdots \gg 2 \gg 1.$$

The next theorem is proven in [21].

Theorem 2.5 (Main Theorem [21]) *Let f be a continuous interval map. We show in if $m \gg s$ and f has a cycle with no block structure of period m then f has also a cycle with no block structure of period s.*

Let us sketch the arguments. We need some observations. Clearly, if a cycle P of rotation pair (k, m) has block structure then k and m are not coprime. Thus, if k and m are coprime then P has no block structure. By Lemma 2.10 we only need to consider convergent cycles. Note that a convergent pattern of rotation pair $(m, 2m)$ and period greater than 2 must have a block structure of special type with two blocks that are swapped by the map; this is because under the circumstances the direction in which a point moves must change on each step and so the entire cycle consists of a block of points to the left of a fixed point and to the right of a fixed point. Thus, a unique pattern with no block structure and rotation number $1/2$ is that of period 2. Since it is forced by any other pattern of period greater than 1 we may consider only patterns of periods greater than 2 and rotation numbers less than $1/2$.

Each integer larger than 2 is of one of the three forms: $2n + 1$, $4n$, $4n + 2$, with $n \geq 1$. The largest possible rotation numbers smaller than $1/2$ for patterns of those periods are, respectively, $\frac{n}{2n+1}$, $\frac{2n-1}{4n}$, $\frac{2n}{4n+2}$. Those numbers are ordered as follows:

$$\cdots < \frac{2n - 1}{4n} < \frac{2n}{4n + 2} = \frac{n}{2n + 1} < \frac{2n + 1}{4n + 4} < \cdots.$$

Let us now prove Theorem 2.5. Using involved arguments, we prove in [21] that a pattern of rotation pair $(2n - 1, 4n)$ forces a pattern of no block structure and rotation pair $(2n, 4n + 2)$. Let us show that, with the above observations, this implies Theorem 2.5. Take a pattern π_1 with no block structure and period $4n$. By Theorem 2.3 it forces a pattern π_2 of rotation pair $(2n - 1, 4n)$. By the above, π_2 forces a pattern with no block structure of rotation pair $(2n, 4n + 2)$. Thus, a no block structure pattern of period $4n$ forces a no block structure pattern of rotation pair

$(2n, 4n + 2)$. If now θ_1 is a no block structure pattern of period $4n + 2$, then by the above it cannot have rotation pair $(2n + 1, 4n + 2)$ which implies (by Theorem 2.3) that its rotation pair forces a pattern θ_2 of rotation pair $(2n, 4n + 2)$, and θ_2, again by Theorem 2.3, forces a pattern θ_3 of rotation pair $(n, 2n + 1)$. Since n and $2n + 1$ are coprime, θ_3 has no block structure. Now, let ξ_1 be a no block structure pattern of period $2n + 1$. By Theorem 2.3 it forces a pattern ξ_2 of rotation pair $(n, 2n + 1)$ which, again by Theorem 2.3, forces a pattern ξ_3 of rotation pair $(2n + 1, 4n + 4)$ which, by the above, has not block structure. This finally proves Theorem 2.5.

In paper [22], these results are used to establish the order among cyclic interval patterns characterized by their *renormalization towers*.

Definition 2.14 (Renormalization tower) Let π be a cyclic permutation of $X = \{1, \ldots, n\}$. Let $p_1 < \cdots < p_s$ be periods of all possible block structures on π. Call the finite string $(p_1/1, p_2/p_1, \ldots, p_s/p_{s-1}, n/p_s)$ the *renormalization tower of* π and denote it $\mathrm{RT}(\pi)$. Similarly we define and denote the renormalization towers of cycles. Let $\gg$ be the lexicographic extension of the order $\gg$ onto renormalization towers.

For brevity by "towers" we always mean "renormalization towers". A tower $\mathscr{N}$ *forces* a tower $\mathscr{M}$ if any pattern of tower $\mathscr{N}$ forces a pattern of tower $\mathscr{M}$ (equivalently, if any map with cycle of tower $\mathscr{N}$ must have a cycle of tower $\mathscr{M}$).

Theorem 2.6 (Main Theorem [22]) *A tower $\mathscr{N}$ forces a tower $\mathscr{M}$ if and only if $\mathscr{N} \gg \mathscr{M}$.*

Let us sketch the proof of Theorem 2.6. Consider the case when $\mathscr{N} = (p_1, \ldots, n/p_s)$ while $\mathscr{M} = (q_1, \ldots, m/q_r)$. Suppose that $p_1 \gg q_1$. Take a pattern π such that $\mathrm{RT}(\pi) = \mathscr{N}$. We need to show that π forces a pattern of tower $\mathscr{M}$. Take a π-canonical map f. It has the cycle of intervals of period p_1, and we can choose a periodic orbit P_1 of period p_1 so that each interval from the cycle contains a unique point from that periodic orbit. Excluding the trivial case when $p_1 = 2$ that can be easily considered separately we conclude that P_1 has not block structure (because by Definition 2.14 we consider *all* possible block structures). Hence by Theorem 2.5 the map f has a cycle Q of period q_1 and no block structure. Moreover, by Theorem 2.4 not only Q but also all extensions of Q by unimodal patterns are exhibited by f. It is not hard to prove (see [22]) that all possible towers are represented among unimodal maps. Hence, the above implies that any tower of the form $(q_1, \ldots)$ is forced by $\mathscr{N}$ as desired. This proves a part of Theorem 2.6 in an important particular case. This should give the reader a good idea as to how Theorem 2.6 is proven in general.

2.4 Coexistence of Homoclinic Trajectories and Stratification of the Space of Maps

The concepts of a *homoclinic trajectory* and a *horseshoe* play a significant role in the theory of dynamical systems. In this section, we deal with them in the context of one dimension. Not surprisingly, in a rather natural way, these concepts lead to the

associated stratification of the space of all continuous interval map related to periods
of cycles and, therefore, to the Sh-theorem.

2.4.1 Homoclinic Trajectories, Horseshoes, and L-Schemes

Along with periodic trajectories, homoclinic trajectories play an important role in the
dynamics. Their presence indicates the presence of trajectories with very complex
behavior. In particular, homoclinic trajectories are thoroughly studied for multidimensional dynamical systems (see [36] and references therein).

Recall that the ω-limit set (of a point x under a map $f : X \to X$ of a topological
space X to itself) is the set of all limit points of the forward f-trajectory of x, i.e.,
the sequence $x, f(x), f^2(x), \dots$. If the map is one to one we can talk about the
backward f-trajectory of x, i.e., the sequence $\dots, f^{-2}(x), f^{-1}(x), x$ and its limit
set called the α-limit set (of x under f). Thus, in the one-to-one case one can also
consider the (well-defined) full trajectory $\dots, f^{-2}(x), f^{-1}(x), x, f(x), f^2(x), \dots$.
In this setting, the full trajectory of x (and x itself if it is not periodic) is said to be
homoclinic if the ω-limit set and the α-limit set of x coincide with the same cycle.

However, if the map is not one to one, the notions of the backward trajectory and
the α-limit set of a point are slightly more complicated. In particular, this relates to
homoclinic trajectories of continuous interval maps because such maps are normally
multivalued. Therefore, the definitions and, to some extent, the terminology need to
be adjusted. For brevity in what follows we will often use notation x_i for a point $f^i(x)$
(if $i \geq 0$) or for an appropriate preimage of x (if i is negative). For example, when
we write x_{-3} we mean a point such that $f^3(x_{-3}) = x_0 = x$. Thus, when we write
$\dots, x_{-2}, x_{-1}, x_0$ we mean simply a branch of the backward trajectory of $x = x_0$.
Now, by a two-sided trajectory of a point $x = x_0$ we mean a two-sided sequence
$\dots, x_{-2}, x_{-1}, x_0, x_1, x_2, \dots$. Evidently, in the setting of not one-to-one maps, a two-
sided trajectory of x_0 is not necessarily unique.

Definition 2.15 Call a two-sided trajectory $\dots, x_{-2}, x_{-1}, x_0, x_1, x_2, \dots$ (and a
point x_0 if x_0 is not periodic) homoclinic (to a cycle $B = (\beta_1, \dots, \beta_m)$), if the ω-limit
set of x_0 is B, and $x_{-jm+k} \to \beta_k$, $k = 1, \dots, m$, as $j \to \infty$.

For brevity from now on when we talk about a homoclinic trajectory, we will
actually mean a **two-sided** homoclinic trajectory. It is natural to expect that the coexistence of homoclinic trajectories is closely related to the coexistence of cycles that
are limits of these homoclinic trajectories. To describe the coexistence of homoclinic
trajectories for one-dimensional dynamical systems, the following classification of
homoclinic trajectories was proposed in [25]; in Definition 2.16, we use notation
from Definition 2.15.

Definition 2.16 A homoclinic trajectory $\dots, x_{-2}, x_{-1}, x_0, x_1, x_2, \dots$ to a cycle B
is said to be one sided if for each k, $k = 1, \dots, m$ the sequence x_{-jm+k} converges
to β_k from one side when $j \to \infty$. A homoclinic trajectory T to a cycle B is said

to be *two sided* if for each k, $k = 1, \ldots, m$ the sequence x_{-jm+k} converges to β_k alternating sides as $j \to \infty$.

We relate the notion of (one-sided) homoclinic trajectory and other notions such as L-scheme and horseshoe (see Definition 2.17).

Definition 2.17 If for a map f there are two **disjoint** closed intervals I and J such that $f(I) \cap f(J) \supset I \cup J$ then we say that f *admits a strict horseshoe* and *intervals I and J form a strict horseshoe for f*. If the intervals I and J **share an endpoint** while we still have $f(I) \cap f(J) \supset I \cup J$ then we say that f *has (admits) a non-strict horseshoe* and *intervals I and J form a non-strict horseshoe for f*. If there exist points $a < x < y$ such that $f(y) \leq a = f(a) < x < y \leq f(x)$ (or points $a > x > y$ such that $f(y) \geq a = f(a) > x > y \geq f(x)$) then f is said to have (to admit) an *L-scheme*.

Remark 2.1 In the original proof of the Sh-theorem, L-scheme was used, in particular, when it was proven that if f has L-scheme then it has cycles of all periods (see Lemma 2.11). Observe that if, say, $x < y < z$, $f(x) < x$, $f(y) \geq z$ and $f(z) \leq x$ then the intervals $[x, y]$ and $[y, z]$ have f-images containing their union, and f has a non-strict horseshoe. This is a prototype of a famous S. Smale horseshoe [37] which was later on discovered in one-dimensional case by Michał Misiurewicz and Wiesław Szlenk [29, 33].

If the map is fixed upfront we may skip mentioning it when talking about its horseshoes. Also, by Remark 2.1 if a map has an L-scheme then it admits a non-strict horseshoe. Clearly, if f has a strict horseshoe then it has a non-strict horseshoe (indeed, if $I < J$ form a strict horseshoe we can extend either interval to a new common point t that belongs to the open interval between I and J). If f has a horseshoe of some sort we simply say that f has a *horseshoe*. To relate these notions further we need the next lemma. Let $\pi(f)$ be the set of all points x such that $f(x) > x$ and $\lambda(f)$ be the set of all points x such that $f(x) < x$.

Lemma 2.15 ([15]) *Suppose that f does not admit L-scheme. Consider a point x such that $\omega(x)$ is not a singleton. Then there exist intervals $(a, c) \subset \pi(f)$ and $(d, b) \subset \lambda(f)$ and a number N such that the following holds:*

1. $a < c \leq d < b$;
2. *for $n \geq N$ we have $f^n(x) \in (a, c) \cup (d, b)$;*
3. $a \notin \omega(x), b \notin \omega(x)$, *and if $c < d$ then also $c \notin \omega(x), d \notin \omega(x)$.*

Proof Clearly, for each n there exists an interval $U \in \pi(f) \cup \lambda(f)$ such that $f^n(x) \in U$. Moreover, if $U = (a', b') \in \pi(f)$ then, by Lemma 2.11, $f^m(x) > a'$ for every $m \geq n$. Let $\{n_k\}$ be the sequence of all numbers such that $f^{n_k}(x) \in U_k = (a_{n_k}, b_{n_k}) \in \pi(f)$. By the above $a_{n_k} \leq a_{n_{k+1}}$ for any k. Let $a = \sup a_{n_k}$.

We claim that there exist an interval $(a, c) \in \pi(f)$ and a number N_1 such that $f^{N_1}(x) \in (a, c)$. Indeed, otherwise if $f^n(x) > a$ for some n then $f^{n+1}(x) < f^n(x)$. On the other hand, by Lemma 2.11 if $f^n(x) < a$ then $f^{n+1}(x) > f^n(x)$. Hence for any k and any $m \geq n_k$ we have $f^m(x) \in [a_{n_k}, \max(f[a_{n_k}, a]]$ which by continuity implies that $\omega(x) = \{a\}$, a contradiction. Hence, $(a, c) \in \pi(f)$ exists.

Similarly, there exist an interval $(d, b) \in \lambda(f)$ and a number N_2 such that $f^{N_2}(x) \in (d, b)$ and for any $n \geq N_2$ we have $f^n(x) < b$ while if $f^n(x) < d$ then $f^{n+1}(x) > f^n(x)$. Combining the two above claims we get that for $N = \max(N_1, N_2)$ and any $n \geq N$ we have $f^n(x) \in (a, c) \cup (d, b)$.

Set $\min \omega(x) = \alpha$, $\min f|_{[d,b] \cap \omega(x)} = \beta$. Clearly, $\alpha \leq \beta$. We claim that $\alpha = \beta$. Indeed, choose $M > N$ so that for any $n \geq M$ we have $d(f^n(x), \omega(x)) \leq \delta$. Then choose $k > M$ with $f^k(x) \in (b, c)$. It follows that if $i > k$ is the first time $f^i(x) \in (a, c)$ then $f^i(x) > \beta - \delta$. Then x maps to the right for some time, enters (d, b), and then, as before, x stays in (d, b) until it maps to (a, c) staying greater than $\beta - \delta$ as before. Thus, for any $j > k$ we have that $f^j(x) > \beta - \delta$. Since δ can be chosen arbitrary, it follows that $\alpha \geq \beta$. Thus, $\alpha = \beta$ as desired. Analogously we can show that $\max \omega(x) = \max f|_{[a,c] \cap \omega(x)}$.

By the previous paragraph we can find a point $y \in (a, c)$ such that $f(y) \in (d, b)$, $f^2(y) = \min \omega(x)$, and so if $\min \omega(x) = a$ this will imply that f has L-scheme, a contradiction. Hence, $a \notin \omega(x)$, and, similarly, $b \notin \omega(x)$. It remains to show that if $c < d$ then $c \notin \omega(x), d \notin \omega(x)$. Indeed, suppose that $c < d$ and $c \in \omega(x)$. Then, since x can never enter $[c, d]$, it follows that if $f^j(x)$ is sufficiently close to c from the left, then from this time on the point x cannot leave $[f^j(x), c]$ and, hence, its trajectory converges to c, a contradiction. □

Lemma 2.15 shows that the absence of L-scheme has far-reaching consequences for the dynamics of the map, as it relates both to its periodic points and to its infinite limit sets. It is used in the next lemma.

Lemma 2.16 *If an interval map f has a horseshoe then f has an L-scheme.*

Proof Let J_0 and J_1 form a horseshoe for f. Assume that J_0 is located to the left of J_1 except, perhaps, for t which is their common endpoint. Let u' (u'') be the leftmost (the rightmost) fixed point in J_0; let v' (v'') be the leftmost (the rightmost) fixed point in J_1. We may assume that $u'' < v'$. Indeed, otherwise $u'' = v' = t$. Choose subintervals J_0' of J_0 and J_1' of J_1 that map onto $J_0 \cup J_1$ so that the endpoints of J_0' and of J_1' map to endpoints of $J_0 \cup J_1$. Since J_0' and J_1' do not contain t then they are disjoint. Replacing J_0 by J_0' and J_1 by J_1' we get the desired. So we may assume that $u'' < v'$; thus, if J_0 and J_1 have a common endpoint t then t is not a fixed point.

A standard pullback construction (see Chap. 1) implies that for any infinite one-sided sequence $\bar{i} = (i_0, i_1, \dots)$ from the space Σ_2 of all sequences with two symbols 0 and 1 there exists a closed interval/point $K_{\bar{i}} = K$ (called an $(\bar{i}\text{-})$fiber) such that $f^r(K) \subset J_{i_r}$ for every r. Choose $\bar{i}$ with dense trajectory in Σ_2 under the one-sided shift σ_2, and then choose x in the $\bar{i}$-fiber. Clearly, $\omega(x)$ is not a fixed point as otherwise, by the above, it belongs to only one of the intervals J_0, J_1 which contradicts the density of the trajectory of $\bar{i}$ in Σ_2.

Assume that f does not have L-scheme and choose fixed points $a < c \leq d < b$ for the point x as in Lemma 2.15. Clearly, $J_0 < c$ is impossible as otherwise, by continuity and Lemma 2.15, the trajectory of x cannot have arbitrarily long strings of iterates that belong to J_0. Similarly, $J_1 > d$ is impossible. This implies that $c < d$ as otherwise $c = d = J_0 \cap J_1$, a contradiction with the previous paragraph. By Lemma 2.15 this implies that $\{a, c, d, b\} \cap \omega(x) = \emptyset$. Again by continuity and by Lemma 2.15 this implies that the trajectory of x cannot have arbitrarily long strings of iterates that belong to J_0 (J_1).   □

We combine all one can say about the relation between different types of horse-shoes and L-scheme in the next lemma.

Lemma 2.17 *An interval map f has an L-scheme if and only if it has a horseshoe. If f has a non-strict horseshoe then f^2 has a strict horseshoe.*

Proof If f has an L-scheme then it has a non-strict horseshoe by Remark 2.1. If f has a horseshoe then it has L-scheme by Lemma 2.16. Hence, it remains to show that if f has a non-strict horseshoe then f^2 has a strict horseshoe.

Assume that I and J form a non-strict horseshoe for f; set $I \cap J = \{x\}$. Consider f^2; then by Lemma 1.4 from Chap. 1 we can choose two subintervals of I such that their f-images are subintervals of I or J, respectively, while their f^2-images equal $I \cup J$. These subintervals of I can be found so that they have at most one point in common. Indeed, their intersection, if non-empty, must be an interval, say, K, and by the assumption $f(K) = x$. Hence, we can simply remove the interior of K from either of the two subintervals we are discussing to get two new subintervals that are disjoint and have the same image as before as desired. Similar choice of subintervals can be made inside J. Choosing subintervals $I' \subset I$, $J' \subset J$ that do not contain x we will see they are *disjoint* and such that $f^2(J') \cap f^2(I') \supset I' \cup J'$.   □

In the next lemma, we relate the existence of a homoclinic trajectory to a fixed point with the existence of an L-scheme.

Lemma 2.18 *An interval map f has a one-sided homoclinic trajectory to a fixed point if and only if it has an L-scheme.*

Proof Suppose that f has a point x with one-sided homoclinic trajectory to a fixed point a. We may assume that $a < x < f(x)$. If there exists a fixed point b with $a < b < x$ then the fact that $\omega(x) = a$ and Lemma 2.11 imply that f has L-scheme. Hence, we may assume that $f(x) > y$ for any $y \in (a, x]$. Suppose that $f^k(x) > a$ for any k. Then the fact that x has to get closer and closer to a implies that in the forward trajectory of x there are longer and longer segments $a < f^i(x) < \cdots < f^j(x)$ with $f^j(x) > x$. Hence, $\omega(x)$ contains points to the right of x, a contradiction. Thus, $f^k(x) \le a$ for some k; by Lemma 2.11 this implies that f has L-scheme.

On the other hand, suppose that f has L-scheme. We may assume that $f(y) \le a = f(a) < x < y \le f(x)$. Replacing a by the closest to x from the left fixed point b we may assume that $f(t) > t$ for any $t \in (a, x]$. Choose a point $z_0 \in (x, y]$ such that $f(z_0) = a$. Then choose a branch of backward trajectory of z_0 contained in (a, z_0) (first choose $z_{-1} \in (a, z)$ with $f(z_{-1}) = z_0$, then $z_{-2} \in (a, z_{-1})$ with $f(z_{-2}) = z_{-1}$, etc.). Clearly, such backward branch exists and converges to a fixed point between a and z_0, i.e., to a. This gives rise to a one-sided homoclinic trajectory to a as desired.   □

Putting together these lemmas we arrive at the next theorem.

Theorem 2.7 *Let f be a continuous interval map. Then f has a one-sided homo-clinic trajectory to a fixed point $\iff$ f has an L-scheme $\iff$ f has a horseshoe.*

Apparently, the concept of *homoclinic trajectory* appeared in one-dimensional dynamics for the first time in [35], where it was observed that a homoclinic trajectory exists in a system if and only if there exists a cycle of period $\neq 2^m$, $m > 0$. Later on, Louis Block [7] proved this claim. We prove a version of the results of [35] and [7] in which L-scheme is involved (below we consider $1 = 2^0$ as a power of 2).

Theorem 2.8 *Let $f \in C^0(I, I)$. Then the following claims are equivalent:*

(1) The map f has a periodic point whose period is not a power of 2.
(2) The map f has a one-sided homoclinic trajectory.
(3) There are disjoint closed intervals J and K in I, and a positive integer n, such that $f^n(J) \supseteq J \cup K$ and $f^n(K) \supseteq J \cup K$.
(4) Some power f^m of the map f has L-scheme.

Proof By Lemma 2.18 the map f^m has L-scheme if and only if f^m has a one-sided homoclinic trajectory. Hence, we only need to show that (1), (3), and (4) are equivalent. Let us show that (1) implies (3). Indeed, by the Sh-theorem (1) implies that there exists m such that $f^{2^m} = g$ has a point x of period 3. We may assume that $x = g^3(x) < g(x) < g^2(x)$. Then $J = [x, g(x)]$ and $K = [g(x), g^2(x)]$ are two closed intervals with disjoint interiors whose images under $g^2 = f^{2^{m+1}}$ cover their union. In other words, g^2 has a non-strict horseshoe; by Lemma 2.17 this means that (1) implies (3). Again by Lemma 2.17 (3) implies (4). Finally, by Remark 2.1 (4) implies (1). $\qquad\square$

Remark 2.2 The notion of the topological entropy [1] plays a crucial role in the dynamical systems theory, including, of course, one-dimensional dynamics. Still, this book is more concerned with coexistence of periods of cycles and other closely related topics. This is why we did not include any treatment of the topological entropy until now. However, in view of Theorem 2.8, we feel compelled to mention the topological entropy, if only in passing. As there are a lot of papers dealing with the entropy in one-dimensional dynamics and giving exact and prioritized references here would be difficult we refer the reader to the book [2] and its Chap. 4.

Indeed, interval maps described in Theorem 2.8 can in fact be characterized by another equivalent condition: these are exactly the maps with positive entropy. In one direction, it is a straightforward corollary of Theorem 2.8 and the following fact: if an interval map admits a non-strict horseshoe then its topological entropy is positive. The converse claim was stated as a conjecture by R. Bowen in his CBMS lectures [24] and verified by M. Misiurewicz and W. Szlenk in [29] and [33]. An alternative proof of the same fact can be based upon Lemma 2.15. Indeed, the fact that a map does not have periodic points of periods not equal to powers of 2 implies that no power of the map has L-scheme. Hence, one can apply Lemma 2.15 first to f, then to f^2, then to f^4, etc. This yields that the map on each infinite limit set is at most two-to-one semiconjugate to a specific transitive shift on the 2-adic compact group. Since the entropy of transitive shifts on compact groups if zero and because of other well-known properties of the entropy it follows that the entropy of f is zero as desired.

2.4.2 Coexistence (of Homoclinic Trajectories) and Its Stability: Powers of Maps with L-Scheme and Homoclinic Trajectories

Various unspecified powers of a map are involved in Theorem 2.8. A more careful studying of properties of powers of the map allows one to specify these powers and yields an interpretation of the Sh-theorem [15]. Take a family $\mathscr{A}$ of interval maps. Say that k *is stronger than* m *in the sense of* $\mathscr{A}$ if the fact that $f^k \in \mathscr{A}$ always implies that $f^m \in \mathscr{A}$; we will also say that this order on integers is *induced by* $\mathscr{A}$. If $\mathscr{L}$ is the family of all continuous interval maps that admit L-scheme then the fact that k is stronger than m in the sense of $\mathscr{L}$ is denoted by $k \triangleright m$. In a slightly different context, the order (**) was considered in [25] and [26] (see Theorem 2.10). Given a map f, denote by $L(f)$ the set of all powers k with $f^k \in \mathscr{L}$.

Theorem 2.9 *The following is the order among integers induced by* $\mathscr{L}$:

$$1 \triangleright 3 \triangleright 5 \triangleright 7 \triangleright 9 \triangleright \ldots \triangleright 2 \cdot 1 \triangleright 2 \cdot 3 \triangleright 2 \cdot 5 \triangleright \ldots \triangleright 2^2 \cdot 1 \triangleright 2^2 \cdot 3 \triangleright 2^2 \cdot 5 \triangleright \ldots \quad (**)$$

Equivalently, the set $L(f)$ *coincides with a tail* $T(n)$ *of the order* (**) *defined by its* $\triangleright$-*greatest number* n. *Conversely, for every* n *there exists a continuous interval map* f *such that* $L(f) = T(n)$. *Moreover, there are also continuous interval maps such that* $f^i \notin \mathscr{L}$ *for every* i.

The difference between (**) and Sh-ordering is that all powers of two "migrate" in (**) in the respective segments of (**), where they become the "strongest". Thus, the "strongest" of all numbers is 1.

Lemma 2.19 ([15]) *Let* $f^n(b) < b < f(b)$. *Then there exists* $c < b$ *such that* $f(c) \neq c$ *while* $f^n(c) = c$. *Moreover, any map* g *sufficiently close to* f *in the* C^0-*sense has a point* $c_g < b$ *such that* $g^n(c_g) \neq c_g$ *while* $g^n(c_g) = c_g$. *In particular, if* $n > 1$ *is odd then* f *and all* C^0-*close to* f *maps must have a point of period* n.

Proof We may assume that f maps $[0, 1]$ to itself. Consider a few cases.

(1) There exists a such that $f(a) = a < b$. We may assume that for any $z \in (a, b]$ we have $z < f(z)$. Choose $y > a$ so close to a that $f^n(y) > y$. Then there exists $c \in (y, b)$ such that $f^n(c) = c$ while $f(c) \neq c$ as desired.
(2) Suppose that for any point $z \in [0, b]$ we have $z < f(z)$. Then if $f^n(0) = 0$ we can set $c = 0$. On the other hand, if $f^n(0) > 0$ then similar to (1) we can find a desired point $c \in (0, b]$.

The definition of C^0-metric implies that if g is C^0-close to f then $g^n(b) < b < g(b)$. This implies the second claim of the lemma.

Finally, if $n > 1$ is odd then by the above c must be of odd period m such that $1 < m \le n$. By the Sh-theorem, f has a point of period n as desired. By the previous paragraph the same holds for all sufficiently close to f in the sense of C^0-metric maps g.   □

Lemma 2.19 is used in the proof of Lemma 2.20.

Lemma 2.20 ([15]) *If $i \ge 1$ then a continuous interval map f has a point of period $2i + 1 \iff f^{2i+1}$ has L-scheme; thus, f has a point of period $2^n(2i + 1) \iff f^{2^n(2i+1)}$ has L-scheme. Moreover, if f has a point of period $2i + 1(i \ge 1)$ then f^2 has L-scheme.*

Proof By Chap. 1 we may assume that f has a periodic point p such that

$$f^{2i}(p) < f^{2i-2}(p) < \cdots < f^2(p) < p = f^{2i+1}(p) < f(p) < \cdots < f^{2i-1}(p).$$

Choose a point $p_{-1} \in [p, f(p)]$ so that $f(p_{-1}) = p$. It follows that $f^{2i+1}(p_{-1}) = f^{2i}(p)$. On the other hand, the fact that $f[p, p_{-1}] \supset [p, p_{-1}]$ implies that there exists a point $q \in (p, p_{-1})$ such that $f^{2i+1}(q) = p_{-1}$. Hence, f^{2i+1} has an L-scheme because $f^{2i}(p), q, p_{-1}$ form an L-scheme for f^{2i+1}. Applying this to $g = f^{2^n}$ we obtain the second claim of the lemma.

To show that f^2 has an L-scheme if f has a point of period $2i + 1$, choose $a \in (f, f(p))$ with $f(a) = a$. Then choose $x \in (f^{2i}(p), f^{2i-2}(p))$ with $f(x) = a$ and $x' \in (f^{2i-3}(p), f^{2i-1}(p))$ with $f(x') = x$. Finally, choose $y \in (a, f^{2i-3}(p))$ with $f^2(y) = x'$. It follows that a, y, x' form an L-scheme for f^2 as desired.

Suppose now that f^{2k+1} has an L-scheme, i.e., $a = f^{2k+1}(a) = f^{2(2k+1)}(x) < x < f^{2k+1}(x)$ for some points a and x. If a is of period $2i + 1$ with $1 \le i \le k$ then by the Sh-theorem f has a point of period $2k + 1$. Assume that $f(a) = a$. Now, if $x < f(x)$ then f has L-scheme by Lemma 2.11. Hence f must have points of all periods by Remark 2.1. Assume that $f(x) < x$. Then by Lemma 2.19 the map f has a point of period $2k + 1$. Thus, if f^{2k+1} has an L-scheme then f has a point of period $2k + 1$.   □

We are ready to prove Theorem 2.9.

Proof of Theorem 2.9. Suppose that $f \in \mathcal{L}$. Then by Remark 2.1 f has a point of period 3. Hence by Lemma 2.20 f^3 has L-scheme and so $f^3 \in \mathcal{L}$. Moreover, if $f^{2k+1} \in \mathcal{L}, k \ge 1$ then by Lemma 2.20 the map f has a point of period $2k + 1$. Hence by the Sh-theorem the map f has a point of period $2k + 3$ which, again by Lemma 2.20, implies that f^{2k+3} has an L-scheme and so $f^{2k+3} \in \mathcal{L}$. This covers the first countable segment of the order $(**)$ from Theorem 2.9.

Now, by Lemma 2.20 the fact that f has a point of an odd period greater than 1 implies that f^2 has an L-scheme, i.e., $f^2 \in \mathcal{L}$. This and the previous paragraph applied to f^2 imply the second countable segment of the order $(**)$ from Theorem 2.9. Repeating these arguments we get the entire order $(**)$ from Theorem 2.9 as desired.

The remaining "realization" part of Theorem 2.9 follows from Lemma 2.20, the realization part of the Sh-theorem, and simple additional arguments. We leave this part of the theorem to the reader.   □

It turns out that the same order $(**)$ controls coexistence of homoclinic trajectories to points of various periods. Recall that we defined one-sided and two-sided homoclinic trajectories in Definition 2.16.

Definition 2.18 We shall call a homoclinic trajectory m-*homoclinic* if it is an one-sided homoclinic trajectory to a cycle of period m or a two-sided homoclinic trajectory to a cycle of period $m/2$.

Thus, if a homoclinic trajectory is m-homoclinic and m is odd then this trajectory is one-sided homoclinic to a cycle of period m, e.g., a 1-homoclinic trajectory is in fact a one-sided homoclinic trajectory to a fixed point.

Theorem 2.10 *If $f \in C^0(I, I)$ has a m-homoclinic trajectory, then f has also a k-homoclinic trajectory for every k such that $m \triangleright k$. Equivalently, the set $H(f)$ of all numbers m such that f is m-homoclinic coincides with a tail $T(n)$ of the order $(**)$ defined by its $\triangleright$-greatest number n. Conversely, for every n there exists a continuous interval map f such that $H(f) = T(n)$. Moreover, there exist continuous interval maps such that $f^i \notin \mathscr{L}$ for every i.*

In other words, forcing among m-homoclinic trajectories induces the order $(**)$ among periods of their limit cycles.

In a few forthcoming lemmas, we deal with a simplest cycle P of odd period m. For brevity we now introduce special notation used concerning P that is used in the proofs of those lemmas. Namely, set $P = \{p_1, p_2, ..., p_m\}$, where $f(p_i) = p_{i+1}$ and $p_m < p_{m-2} < \cdots < p_1 < p_2 < \cdots < p_{m-1}$. Since $f(p_1) > p_1$ and $f(p_2) < p_2$, then there is an f-fixed point $a \in [p_1, p_2]$. Let $I_1 = [p_1, a]$, $I_2 = [a, p_2]$, $I_3 = [p_3, p_1]$, $I_4 = [p_2, p_4]$, ..., $I_{m-1} = [p_{m-3}, p_{m-1}]$, $I_m = [p_m, p_{m-2}]$. Clearly, $f(I_j) \supset I_{j+1}, 1 \le j \le m$ and $f(I_m) \supset I_1$. Hence, $\mathscr{I} = I_1 \to I_2 \to \cdots \to I_1$ is a loop of intervals.

Just in case let us remind the reader a few previously introduced concepts.

Definition 2.19 (Block structure) Let $X = \{x_1 < \cdots < x_n\}$ be a cycle of a map f. Suppose that for some $k > 1, m > 1$ we have $n = km$ and the map f maps sets $Y_1 = \{x_1, \ldots, x_m\}$, $Y_2 = \{x_{m+1}, \ldots, x_{2m}\}$, ..., $Y_k = \{x_{n-m+1}, \ldots, x_n\}$ to one another. Then sets $Y_1, \ldots, Y_k$ are called *blocks (of X)* and the cycle X is said to have *block structure*; if blocks are two-point sets, X is said to be a *doubling*. Otherwise X is said to have *no block structure*.

Now we are ready to state an important and deep result from [2] which specifies the concept of a simplest cycle of period $2^n(2k + 1)$ where $k, n \ge 1$.

Theorem 2.11 (Forcing-minimal cycles of arbitrary period) *Let $m = 2^n(2k + 1)$. If $n = 0$ or $k = 0$ then the forcing-minimal patterns of period m are exactly the simplest patterns. However if $n \geq 1$ and $k \geq 1$ then the forcing-minimal patterns of period m are exactly the simplest patterns that satisfy the following extra conditions: the map from block to block is monotone except for one such map which is unimodal.*

For convenience by $\langle a, b \rangle$, we mean a closed interval with endpoints a, b but without assuming that $a < b$. We need the following technical lemma.

Lemma 2.21 *Suppose that x is a point of a cycle P of period n of a continuous interval map f. Let $y_0, y_1, \ldots, y_{n-1}$ be points such that if we set $I_j = \langle f^j(x), y_j \rangle$, then $I_j \subset \langle f^j(x), f(y_{j-1}) \rangle$ for each $j = 1, \ldots, n - 1$, $\langle x, y_0 \rangle \subset \langle x, f(y_{n-1}) \rangle$, and $f(y_i) \neq y_{i+1}$ for some i. Then there exists a point $z \in I_0$ such that*

1. *$f^j(z) \in I_j, 0 \leq j \leq n - 1, f^n(z) = z$;*
2. *there exist intervals $K_0 = \langle z, y_0 \rangle \supset K_{-1} \supset \ldots$ that all have z as an endpoint, whose intersection is $\{z\}$, and such that $f^j(K_{-1}) \subset \langle f^j(z), y_j \rangle, j = 0, \ldots, n - 1$ and $f^n(K_{-r}) = K_{-r+1}$ for any $r \geq 1$.*

Proof Clearly, $\mathscr{I} = I_0 \rightarrow I_1 \rightarrow \cdots \rightarrow I_{n-1} \rightarrow I_0$ is a loop of intervals. If we pull back I_0 along $\mathscr{I}$, we will after n steps obtain an interval $I_0^1 \subset I_0$ that follows $\mathscr{I}$ and is such that $f^n(I_0^1) = I_0$. Since by the assumptions $f(y_i) \neq y_{i+1}$ for some i, then $I_0^1 \neq I_0$. However, a slight shrinkage of I_0 that follows from that does not necessarily imply that after countably many steps I_0 will shrink to x. In fact, there may be an interval with an endpoint x in I_0 on which f^n is the identity; then this interval will be always contained in any pull back of I_0 along $\mathscr{I}$. To bypass this problem, choose the point $z \in I_0$ which is the farthest from x point that is f^n-fixed and follows $\mathscr{I}$. If we replace x by z, we will see that literally the same properties that were required originally hold for intervals $I_j' = \langle f^j(z), y_j \rangle$; the loop of these intervals will be denoted $\mathscr{I}'$. The choice of z implies that as we pull back $I_0' = K_0$ along $\mathscr{I}'$ we will get intervals $K_0 \supset K_{-1} \supset \ldots$ with an endpoint z that shrink to z. □

We are ready to prove the next lemma that establishes the relation between some cycles and homoclinic trajectories.

Lemma 2.22 *If P is a cycle of $f \in C^0(I, I)$ of period n which is not a doubling then f has a one-sided homoclinic trajectory to a cycle of the same period as P. In particular, if f has a point of period m which is not a power of 2 then f has a one-sided homoclinic trajectory to a cycle of period m.*

Proof Let $L = [a, b]$ be the leftmost P-basic interval. By definition intervals from $\mathscr{F}_P$ satisfy all conditions of Lemma 2.21 except that claiming the existence of i with $f(y_i) \neq y_{i+1}$. To see that such i exists, assume otherwise. Denote by g the P-linear map. Then $\langle g^i(a), g^i(b) \rangle$ is a P-basic interval for all i, $0 \leq i \leq n$. In particular if $g^k(b) = a$, then $g^k(a) = b$. Hence the first k steps of the g-trajectory of L will then be repeated for k more steps with points a and b "flipped" so that $n = 2k$. These k intervals are pairwise disjoint as otherwise L itself is non-disjoint from, say, $g^r(L) \neq L$ implying that $g^r(a) = b$, and hence $r = k$ and $g^r(L) = g^k(L) = L$, a contradiction. Thus, P is a doubling, a contradiction. Hence there exists, say, the minimal i such that $f^i(b)$ is not equal to the endpoint of the i-th interval of $\mathscr{F}_P$ distinct from $f^i(a)$.

By Lemma 2.21 there exists $x = f^n(x) \in L$ that follows $\mathscr{F}_P$ and is such that there are no other such points in $(x, b]$. We claim that x is of period n. Suppose otherwise. Then x is of period $k < n$. Since each P-basic interval can be repeated in $\mathscr{F}_P$ at most twice (one time for each moment when a maps to its endpoints), the only way k may be less than n is when $n = 2k$. Hence, there are k intervals in $\mathscr{F}_P$ and each is repeated in $\mathscr{F}_P$ twice. Since x each such an interval with the delay of k steps, the endpoints of this interval are flipped under g^k. Hence, these intervals are mapped onto one another by g. As in the previous paragraph, this implies that P is a doubling, a contradiction.

By Lemma 2.21 there exists a nested sequence of intervals with an endpoint z and intersection $\{z\}$ which are pullbacks of $[x, b]$. If we now map $[x, b]$ forward, we will, at the moment $i + 1$, cover completely a P-basic interval J not containing $f^{i+1}(x)$ (the number i was found in the first paragraph of the proof). Since this interval at some moment f-covers L, then it contains preimages of x not equal to $f^{i+1}(x)$. If we pull back this interval to $[x, b]$, we will see that $[x, b]$ contains a preimage $u \neq x$ of x. If we now use Lemma 2.21 to construct the backward trajectory of u, we will finally arrive at a full trajectory of u which is homoclinic to the orbit of x; by construction this full trajectory is one-sided homoclinic as desired. To prove the last claim of the lemma notice that by Lemma 2.21 if f has a point of period $m = 2^n(2k + 1), k \geq 1$, then it has a cycle of period m which is not a doubling. $\square$

Let us recall a few standard definitions.

Definition 2.20 (Markov graph) Let $P = \{x_1 < x_2 < \cdots < x_n\}$ be a cycle of an interval map f. Then each interval $[x_i, x_{i+1}]$ is said to be P-*basic*. An P-*linear* map g coincides with f on P and is linear on each P-basic interval. The Markov graph M_P of P has vertices that are the P-basic intervals and arrows correspond to g-*covering* (there is an arrow from J to K if $K \subset g(J)$). The loops in this graph correspond to cycles of g and, actually, f.

Some loops in M_P are more important than the others. In particular, the fundamental loop of intervals defined below is used for finding homoclinic trajectories.

Definition 2.21 (Fundamental loops) If P is a cycle of period $n > 1$ then for each point $x \in P$ we consider *germs* at x, i.e., small intervals with x as one of the endpoints. Each point of P has two germs, except the leftmost and rightmost points, which have one germ each. There is a natural map induced on the set of germs by the P-linear map f, and if we start at the germ of the leftmost point, we pass through the germ of the rightmost point and get back exactly after n applications of this map. This loop of germs is called the *fundamental loop of germs (of P)*. Each germ is contained in a unique P-basic interval, so this loop of germs gives us a loop $\mathscr{F}_P$ of P-basic intervals called *fundamental loop of (P-basic) intervals*.

Unlike Lemma 2.22, the next lemma shows that in some cases cycles of certain periods force homoclinic trajectories of distinct periods.

Lemma 2.23 *Let $f \in C^0(I, I)$ have a periodic point of period $m = 2^n(2k + 1), k \geq 1$. Then f has a 2^{n+1}-homoclinic trajectory.*

Proof By Theorem 2.11 we can choose a cycle P of f of period m which is forcing-minimal. Set $s = 2^n - 1$ and denote by $P_0, \ldots, P_s$ blocks of P such that $P_i = f^i(P_0), 0 \leq i \leq s$. There exists a cycle $\{a_0, \ldots, a_s\}$ such that $f^i(a_0)$ belongs to the convex hull J_i of $P_i, 0 \leq i \leq s$, $f^s(a_0) = a_0$, and for each a_i the endpoints of the P-basic interval containing a_i map in distinct directions under f^s. Denote by I_r, J_r the two intervals into which a_0 partitions the P-basic interval containing $a_r, 0 \leq r \leq s$. We may chose the notation so that the following is a loop of intervals of length $2s + 2$:

$$\mathscr{I} = I_0 \to I_1 \to \ldots I_s \to J_0 \to J_1 \to \ldots J_s \to I_0.$$

Evidently, this loop of intervals satisfies all the conditions of Lemma 2.21. Hence there exists a point $x \in I_0$ satisfying all the conclusions of Lemma 2.21. We may assume that $I_0 = [p_0, a_0], J_0 = [a_0, p_1]$ where $f^{s+1}(p_0) = p_1$; clearly, $f^{2^{n+1}}(x) = x$. Moreover, the structure of P implies that some eventual images of (p_0, x) contain x. Pulling back $[p_0, x]$ along $\mathscr{I}$, we, similarly to Lemma 2.22, arrive at the full homoclinic trajectory to the (periodic) orbit X of x. There are two possibilities here. First, it may be that $x \neq a_0$. Then the point x is of true period 2^{n+1} and the just found homoclinic trajectory is one sided to X. Second, it may happen so that $x = a_0$. Then x is of period 2^n and the just found homoclinic trajectory is two sided to X. In either case, this homoclinic trajectory is 2^{n+1}-homoclinic as desired.   □

We are ready to prove Theorem 2.10.

Proof of Theorem 2.10. By definition, any 1-homoclinic trajectory is a one-sided homoclinic trajectory to a fixed point. By Lemma 2.18, this implies that if f has a 1-homoclinic trajectory then f has an L-scheme. By Lemma 2.11 f has points of all periods. In particular, f has a point of period 3. Hence by Lemma 2.22 f has a one-sided homoclinic trajectory to a cycle of period 3. In other words, the existence of a 1-homoclinic trajectory implies the existence of a one-sided homoclinic trajectory to a cycle of period 3.

Now, suppose that f has an m-homoclinic trajectory (here $m = 2^n(2k + 1)$). Then either it is a one-sided homoclinic trajectory to a cycle of period m or a two-sided homoclinic trajectory to a cycle of period $m/2$. By the Sh-theorem in either case f has a cycle of period m. Then, again by the Sh-theorem, f has a cycle of period $m + 2$. By Lemma 2.22 f has a one-sided homoclinic trajectory to a cycle of period $m + 2$. In other words, the existence of an m-homoclinic trajectory implies the existence of a one-sided homoclinic trajectory to a cycle of period $m + 2$.

By Lemma 2.23, if f has an m-homoclinic trajectory (here $m = 2^n(2k + 1)$), then f has a 2^{n+1}-homoclinic trajectory.

It remains to show that if f has a 2^{n+1}-homoclinic trajectory then it has a $2^{n+1}3$-homoclinic trajectory. Indeed, the fact that f has a 2^{n+1}-homoclinic trajectory implies that $g = f^{2^{n+1}}$ has a one-sided homoclinic trajectory to a g-fixed point. By Lemma 2.11 g has points of all periods. In particular, g has a point x of period 3. Since $f^{2^{n+1}3}(x) = x$ while $f^{2^{n+1}}(x) \neq x$, the period of x under f is $2^i 3$ with $i \leq n + 1$. By the Sh-theorem this implies that f has a cycle of period exactly $2^{n+1}3$. By Lemma 2.22 f has a one-sided homoclinic trajectory to a cycle of period $2^{n+1}3$.

The remaining "realization" part of the theorem is left to the reader.   □

The Sh-theorem, Theorems 2.9 and 2.10 can be interpreted as results establishing relations of containment among certain subsets of the space $C^0(I, I)$ of continuous interval maps.

Definition 2.22 Let $F(m)$ be the set of all $f \in C^0(I, I)$ with cycle of period m. Let $L(m)$ be the set of all $f \in C^0(I, I)$ such that f^n has an L-scheme. Let $H(m)$ be the set of all $f \in C^0(I, I)$ having a m-homoclinic trajectory.

The results of this section show that these sets are closely related. Indeed, by Lemmas 2.22 and 2.20 $F(m) = B(m) = H(m)$ for any integer m which is not a power of 2. By Theorems 2.7 and 2.10 $L(m) = H(m)$ if m is a power of 2. Thus, in particular, $L(m) = H(m)$ for every m. Moreover, the Sh-theorem and our results also imply a series of inclusions among these families of maps. Indeed, by the Sh-theorem

$$F(1) = C^0(I, I) \supset F(2) \supset F(2^2) \cdots \supset F(2^\infty) \cdots \supset F(5) \supset F(3),$$

where by F_{2^∞} we mean the family of maps that have cycles of periods that are all powers of 2; this series of inclusions can be widened based upon the results of this section resulting in the following inclusions:

$$F_1 \supset F_2 \supset F_{2^2} \supset \cdots \supset F_{2^\infty} \supset$$

· · · · · · · · · · · ·

$$\cdots \supset F(5 \cdot 2) = H(5 \cdot 2) = L(5 \cdot 2) \supset F(3 \cdot 2) = H(3 \cdot 2) = L(3 \cdot 2) \supset H(2) = L(2) \supset$$

$$\cdots \supset F(5) = H(5) = L(5) \supset F(3) = H(3) = L(3) \supset H(1) = L(1).$$

Equivalently, these inclusions are stated in the next theorem which can be viewed as a restatement of the Sh-theorem and some results of this section.

Theorem 2.12 *We have that $F(i) \subset F(j)$ if and only if $i \succ j$. Also, $L(k) \subset L(l)$ if and only if $k \rhd l$. Finally, $H(k) \subset H(l)$ if and only if $k \rhd l$.*

So far we have considered coexistence of periods of cycles (including cycles to which homoclinic trajectories converge) and various types of horseshoes for an individual map. This leads to a natural question concerning whether the associated properties of a map $f \in C^0(I, I)$ (such as to have a cycle of period n, or an n-homoclinic trajectory, or to be such that f^n has an L-scheme) are stable under perturbations. To answer this question we prove the following theorem.

Theorem 2.13 *The following claims hold:*

1. *If $k \succ l$ then every map f that has a cycle of period k has a C^0-neighborhood $\mathcal{U}$ such that any map $g \in \mathcal{U}$ has a cycle of period l. In other words, $F_k \in \text{int}(F_l)$.*
2. *If $k \triangleright l$ then every map f such that f^k has an L-scheme has a C^0-neighborhood $\mathcal{V}$ such that any map $g \in \mathcal{V}$ is such that g^l has an L-scheme. In other words, $B_k \in \text{int}(B_l)$.*
3. *If $k \triangleright l$ then every map f such that f has a k-homoclinic trajectory a C^0-neighborhood $\mathcal{V}$ such that any map $g \in \mathcal{V}$ has an l-homoclinic trajectory. In other words, $H_k \in \text{int}(H_l)$.*

Part (1) of the theorem was first proven by L. Block in [9].

Proof (1) Consider first the case when $k > 1$ is odd. By the proven in Chap. 1 we may assume that f has periodic point p such that

$$f^{2i}(p) < f^{2i-2}(p) < \cdots < f^2(p) < p = f^{2i+1}(p) < f(p) < \cdots < f^{2i-1}(p).$$

Choose a fixed point $a \in (p, f(p))$. Then choose a point $y \in (a, f(p))$ such that $f(y) = p$. Then choose a point $x \in (p, a)$ such that $f(x) = y$. It follows that $f^{2p+3}(x) = p < x < f(x)$. Hence, by Lemma 2.19, all maps g sufficiently C^0-close to f have a cycle of period $2p + 3$ as desired. Clearly, this implies the claim of the theorem for all integer periods except for powers of 2.

It remains to show that if a map f has a cycle of period 2^n then all sufficiently C^0-close to f maps g have cycles of period 2^{n-1}. First consider a simplest cycle P of period 4 of a map f. We may assume that $P = \{p_2 < p_0 = p < p_1 < p_3\}$ and $f(p_i) = p_{i+1}, i = 0, 1, 2$ while $f(p_3) = p_0$. If g is C^0-close to f then $g^2(p) < p < g(p) < g^3(p)$ and $g^4(p)$ is so close to p that also $g^2(p) < g^4(p) < g(p) < g^3(p)$. It follows that g^2 maps points p and $g^2(p)$ in opposite directions so that there exists a g^2-fixed point $x \in [g^2(p), p]$. If x is of g-period 2, then we are done. If x is g-fixed then $g^2(p) < x = g(x) < p < g(p)$ is an L-scheme and, hence, g has a point of period 2. We conclude that if $f \in F(4)$ then $f \text{int}(F_4)$. If now $f \in F(2^n), n \geq 2$ then $g = f^{2^{n-2}}$ has a cycle of period 4. By the above a map g sufficiently C^0-close to f is such that $g^{2^{n-2}}$ has a cycle of period 2, and, hence, g itself has a cycle of period 2^{n-1}. In other words, $F(2^n) \subset \text{int}(F(2^{n-1}))$ as desired.

(2) If f has an L-scheme then we may assume that there points $a < x < y$ such that $f(y) \leq a = f(a) < x < y \leq f(x)$ and there are no fixed points between a and x. Choosing preimages we can find a point $z \in (a, x)$ such that $z < f(z) < f^2(z) = y$. Then for all maps g sufficiently C^0-close to f we have $f^3(z) \leq z < g(z) < g^2(z)$.

By Lemma 2.19 this implies that g has a cycle of period 3. By Lemma 2.20 this implies that g^3 has an L-scheme as desired.

Now, by Lemma 2.20 and (1) above we see that if f^{2k+1} has an L-scheme then g^{2k+3} has an L-scheme if g is sufficiently C^0-close to f. This takes care of the first countable segment in the order $(**)$. On the other hand, if f^{2i+1} has an L-scheme then by the just proven all sufficiently C^0-close to f maps g have points of period $2i + 3$ and hence, again by Lemma 2.20, are such that g^2 has an L-scheme. Then we can apply the above arguments, to f^2 and g^2, and so on. In the end, this proves the second claim of the theorem.

(3) If f has a 1-homoclinic trajectory then by Lemma 2.18 it has an L-scheme. Hence it has a point x such that $f^4(x) = f^3(x) < x < f(x) < f^2(x)$. If g is C^0-close to f then $g^3(x) < x < g(x) < g^2(x)$. By Lemma 2.19 g has a point of period 3. By Theorem 2.10, g has a 3-homoclinic trajectory as desired. Now, for an odd number m the fact that f has a point of period m is equivalent to the fact that f has a one-sided homoclinic trajectory to a cycle of period m. Since by (1) a map g close to f must have a cycle of period $m + 2$, we obtain the claim of (3) as it applies to the first countable segment of $(**)$. Now, if f has a 2-homoclinic trajectory then, by definition, f^2 has a one-sided homoclinic trajectory to an f^2-fixed point. Applying to f^2 the just established facts concerning the first countable segment of the order $(**)$ and homoclinic trajectories, we can take care of the second countable segment of $(**)$, and so on. In the end we prove (3) as desired. $\qquad\square$

References

1. Adler, R., Konheim, A., McAndrew, M.: Topological entropy. Trans. Amer. Math. Soc. **114**, 309–319 (1965)
2. Alsedà, L., Llibre, J., Misiurewicz, M.: Combinatonal dynamics and entropy in dimension one. In: Advanced Series in Nonlinear Dynamics, vol. 5, 2nd edn. World Scientific, River Edge, NJ (2000)
3. Alsedà, L., Llibre, J., Misiurewicz, M.: Periodic orbits of maps of Y. Trans. Amer. Math. Soc. **313**, 475–538 (1989)
4. Baldwin, S.: Generalizations of a theorem of Sarkovskii on orbits of continuous real-valued functions. Discrete Math. **67**, 111–127 (1987)
5. Barge, M.: Horseshoe maps and inverse limits. Pacific J. Math. **121**, 29–39 (1986)
6. Benedicks, M., Carleson, L.: The dynamics of the Hénon map. Ann. Math. **133**, 73–169 (1991)
7. Block, L.: Homoclinic points of maps of the interval. Proc. Amer. Math. Soc. **72**, 576–580 (1978)
8. Block, L.: Simple periodic orbits of mappings of the interval. Trans. Amer. Math. Soc. **254**, 391–398 (1979)
9. Block, L.: Stability of periodic orbits in the theorem of Šarkovskii. Proc. Amer. Math. Soc. **81**, 333–336 (1981)
10. Block, L., Coppel, W.A.: Stratification of continuous maps of an interval. Trans. Amer. Math. Soc. **297**, 587–604 (1986)
11. Block, L., Guckenheimer, J., Misiurewicz, M., Young, L.-S.: Periodic points and topological entropy of one dimensional maps. In: Global Theory of Dynamical Systems. Lecture Notes in Mathematics, vol. 819, pp. 18–34. Springer, Heidelberg (1980)

12. Block, L., Hart, D.: The bifurcation of periodic orbits of one-dimensional maps. Ergod. Th. & Dynam. Syst. **2**, 125–129 (1982)
13. Block, L., Hart, D.: The bifurcation of homoclinic orbits of maps of the interval. Ergod. Th. & Dynam. Syst. **2**, 131–138 (1982)
14. Block, L., Hart, D.: Stratification of the space of unimodal interval maps. Ergod. Th. & Dynam. Syst. **3**, 533–539 (1983)
15. Blokh, A. M.: An interpretation of a theorem of A.N. Sharkovsky (in Russian). In: Oscillation and stability of solutions of functional-differential equations. pp. 3–8. Inst. Mat. Akad. Nauk Ukrain.SSR, Kyiv (1982)
16. Blokh, A.M.: Rotation numbers, twists and a sharkovskii-misiurewicz-type ordering for patterns on the interval. Ergod. Th. & Dynam. Syst. **15**, 1–14 (1995)
17. Blokh, A.M.: Functional rotation numbers for one-dimensional maps. Trans. Amer. Math. Soc. **347**, 499–514 (1995)
18. Blokh, A.M.: On rotation intervals for interval maps. Nonlineraity **7**, 1395–1417 (1994)
19. Blokh, A.M., Misiurewicz, M.: New order for periodic orbits of interval maps. Ergod. Th. & Dynam. Syst. **17**, 565–574 (1997)
20. Blokh, A.M., Misiurewicz, M.: Rotation numbers for certain maps of an n-od. Topol. Appl. **114**, 27–48 (2001)
21. Blokh, A.M., Misiurewicz, M.: Forcing among patterns with no block structure. Topol. Proceed. **54**, 125–137 (2019)
22. Blokh, A.M., Misiurewicz, M.: Renormalization towers and their forcing. Trans. Amer. Math. Soc. **372**, 8933–8953 (2019)
23. Bobok, J., Kuchta, M.: X-minimal patterns and a generalization of Sharkovskii's theorem. Fund. Math. **156**, 33–66 (1998)
24. Bowen, R. On Axiom a diffeomorphisms. In: Regional Conference Series in Mathematics. Conference Board of the Mathematical Sciences, vol. 35. Providence (1978)
25. Fedorenko, V. V., Sharkovsky, A. N.: On coexistence of periodic and homoclinic trajectories (in Russian). In: The fifth All-Union Conference on Qualitative Theory of Differential Equations. Abstracts, pp. 74–175. Kishinev (1979)
26. Fedorenko, V. V., Sharkovsky, A. N.: Stability of dynamical system property to have a homoclinic trajectory (in Russian). In: Oscillation and stability of solutions of functional-differential equations, pp. 111–113. Inst. Mat. Akad. Nauk Ukrain.SSR, Kyiv (1982)
27. Hall, T.: Fat one-dimensional representatives of pseudo-Anosov isotopy classes with minimal periodic orbit structure. Nonlinearity **7**, 367–384 (1994)
28. Li, T.-Y., Misiurewicz, M., Pianigiani, G., Yorke, J.: No division implies chaos. Trans. Amer. Math. Soc. **273**, 191–199 (1982)
29. Misiurewicz, M.: Horseshoes for mappings of an interval. Bull. Acad. Pol. Sci., Sér. Sci. Math. **27**, 167–169 (1979)
30. Misiurewicz, M.: Periodic points of maps of degree one of a circle. Ergod. Th. & Dynam. Syst. **2**, 221–227 (1982)
31. Misiurewicz, M.: Formalism for studying periodic orbits of one dimensional maps. In: European Conference on Iteration Theory (ECIT 87), pp. 1–7. World Scientific, Singapore (1989)
32. Misiurewicz, M., Nitecki, Z.: Combinatorial Patterns for Maps of the Interval. Mem. Amer. Math. Soc. **456** (1990)
33. Misiurewicz, M., Szlenk, W.: Entropy of piecewise monotone mappings. Studia Mathematica **67**, 45–63 (1980)
34. Sharkovsky, A.N.: Coexistence of cycles of a continuous mapping of the line into itself (in Russian). Ukrain. Math. Zh. **16**, 61–71 (1964)
35. Sharkovsky, A. N.: About an isomophism problem of dynamical systems. In: Proceedings V International Conference on Nonlinear Oscillations, vol. 2, pp. 541–545. Naukova dumka, Kiev (1970)
36. Gonchenko S.V., Shilnikov L.P. (eds). Homoclinic tangencies. SC "Regular and chaotic dynamics", Inst. Computer Studies. Moscow-Izhevsk (2007)
37. Smale, S.: Differentiable dynamical systems. Bull. of the Amer. Math. Soc. **73**, 747–817 (1967)
38. Ziemian, K.: Rotation sets for subshifts of finite type. Fundam. Math. **146**, 189–201 (1995)

Chapter 3
Coexistence of Cycles for One-Dimensional Spaces

3.1 Circle Maps

Another class of one-dimensional maps for which a full description of the sets of periods of their periodic points is possible is the class of circle maps of degree one. The main results here describe the periods of periodic orbits of circle maps. For the circle maps of degree other than one, they are mainly due to Efremova [20], and Block, Guckenheimer, Misiurewicz, and Young [9]. For the circle maps of degree one, they are due to MichałMisiurewicz [23, 24]. What makes it more interesting is that in this case the situation is quite different from that on the interval.

Indeed, by the Forcing Sh-Theorem periods of periodic orbits of interval maps force each other. For instance, once we know that an interval map f has a periodic orbit of period 3, we are guaranteed that there will be f-periodic orbits of all periods. More generally, once we know that an interval map as a periodic orbit of period n we are guaranteed that there will be f-periodic orbits of all periods m such that $n \succ m$.

In the case of the circle maps of degree one though the information about *only one* periodic orbit is insufficient for making conclusions about other periodic orbits. This is because of the existence of circle rotations (all of whom are maps of degree one). Indeed, denote by S^1 the unit circle. If $f : S^1 \to S^1$ is a degree one circle map that has a point of period m we can simply choose an rotation o S^1 by an angle $\frac{i}{m}$ where i and m are coprime to see that period m *alone* does not automatically force any other periods.

An important difference that to some extent explains the drastic contrast with interval maps can be observed if we consider the following issue. First, we need the following notation: For two points $a, b \in S^1$ let us denote by $[a, b]$ the circle arc with endpoints a to b such that movement in $[a, b]$ in the positive direction takes place from a to b. Now, if $x, y \in S^1$ are two distinct points and $f : S^1 \to S^1$ is continuous then we are only guaranteed that $f([x, y])$ covers one of the two arcs (either $[f(x), f(y)]$ or $[f(y), f(x)]$) connecting $f(x)$ and $f(y)$ in S^1. This is of

A. M. Blokh and O. M. Sharkovsky, *Sharkovsky Ordering*,
SpringerBriefs in Mathematics, https://doi.org/10.1007/978-3-030-99125-8_3

course quite different from the situation on the interval where there is a unique arc connecting any two points.

Another heuristic reason explaining the difference between the dynamics on circle and on the interval is that circle maps do not have any *a priori* given set of periods that their periodic orbits must have. On the other hand, interval maps, unlike circle maps of degree one, must have a fixed point. Thus, on the interval when considering one given periodic orbit of some period we in fact consider this periodic and at least one fixed point. While this is of course not a precise mathematical argument or explanation it offers a guess toward a kind of assumption one may need in order to be able to draw meaningful conclusions concerning sets of periods of periodic points of circle maps of degree one: Here, we need to talk about two periodic orbits.

Moreover, in the case of circle degree one maps another characteristic of periodic orbits seems to be more natural than the period. We mean so-called *rotation numbers* and *rotation pairs* of periodic points describing dynamics of periodic orbits in more detail than just periods. In the end, we will describe sets of rotation numbers and pairs of continuous circle maps of degree one. In our presentation, we follow [2] with some obvious changes (in particular, in this section, we do not include proofs of statements).

We need a few definitions. Let $f : S^1 \to S^1$ be a continuous map. A very important characteristic of the map f is its *degree* normally denoted by $\deg(f)$. Loosely, it shows how many times the image point $f(x)$ rotates around S^1 as x rotates around S^1 once in the positive (i.e., counterclockwise) direction; observe that as it happens, $f(x)$ may well rotate in the positive or negative direction (accordingly, $\deg(f)$ can be positive or negative). An equivalent heuristic description of the degree is that it shows the minimal number of times the image of S^1 covers S^1 in the positive (negative) direction.

A formal definition of the degree is based upon the notion of the *lifting* of f to the real line $\mathbb{R}$. Namely, let $e : \mathbb{R} \to S^1$ be the natural *exponential projection* defined by $e(X) = e^{2\pi i X}$. Then a lifting of f to $\mathbb{R}$ is a map $F : \mathbb{R} \to \mathbb{R}$ such that the exponential projection e semiconjugates F and f. In other words, we have to have $e \circ F = f \circ e$.

Let us fix a lifting $F : \mathbb{R} \to \mathbb{R}$ of f to $\mathbb{R}$. The fact that f is of degree one is equivalent to the identity $F(X + 1) = F(X) + 1$. This implies that $F(X + k) = F(X) + k$ for any integer k. We will rely upon basic properties of liftings and the notion of the degree. In particular, any two liftings of the same circle map are integer shifts of each other, and any integer shift of a lifting of a circle map is a lifting of the same circle map too. Moreover, the if $f : S^1 \to S^1$ and $g : S^1 \to S^1$ are two circle maps then $\deg(f \circ g) = \deg(f) \cdot \deg(g)$ so that, for example, $\deg(f^n) = [\deg(f)]^n$. In particular, any iteration of a circle degree one map is again a degree one circle map.

Before moving on to the most interesting and complicated case of circle maps of degree one, let us sort out the remaining cases.

Theorem 3.1 ([8, 9, 20]) *Let f be a circle map of degree $d \neq 1$.*

1. If $|d| > 1$, and $d \neq -2$, then the set of periods of f is $\mathbb{N}$.

2. *If $d = -2$, then the set of periods of f is either $\mathbb{N}$ or $\mathbb{N} \setminus \{2\}$, and both possibilities can occur.*
3. *If $d = 0$ or $d = -1$, then the set of periods of f is an initial segment of Sh-ordering, and each such an initial segment can be realized (for each of $d = 0$ and $d = -1$).*

For circle maps of degree one, we consider maps of $\mathbb{R}$ of degree one. These are continuous maps $F : \mathbb{R} \to \mathbb{R}$ such that $F(X + 1) = F(X) + 1$ (in the rest of this section, we reserve the notation F for a continuous map $\mathbb{R} \to \mathbb{R}$ of the real line of degree one). As was discussed above, every lifting of a circle degree one map has this property. On the other hand, if $F : \mathbb{R} \to \mathbb{R}$ is a continuous map of degree one then there exists a continuous circle map of degree one $f : S^1 \to S^1$ such that $e \circ F = f \circ e$. Hence, it makes sense to consider maps of the real line of degree one and then transfer the results established for such maps down to circle maps of degree one. Continuous maps of the real line of degree one form a Banach space in sup-metric. We shall always use sup-metric for them. A major tool used in studying the real line maps F of degree one is the construction of their lower and upper versions denoted by F_l and F_u, respectively. They are defined as follows:

$$F_l(x) = \inf\{F(y) : y \geq x\}, \quad F_u(x) = \sup\{F(y) : y \leq x\}.$$

Let us suggest a nice way of describing the map F_u. Imagine that water is poured on the graph of F from above. Then it will fill up all the holes in the graph after which will start pouring out. Hence while all the holes will be completely filled up, there will be no extra water added to the filled-up holes. Similarly one can interpret F_l, associated in the same fashion with the process of pouring water on the graph of F "from below".

Some basic properties of the maps F_u and F_l are listed below; when we write $F \leq G$ we mean that $F(x) \leq G(x)$ for any $x \in \mathbb{R}$.

1. $F_l \leq F \leq F_u$.
2. If $F \leq G$ where $G : \mathbb{R} \to \mathbb{R}$ is a continuous map of degree one then $F_l \leq G_l$ and $F_u \leq G_u$.
3. If F is non-decreasing then $F = F_l = F_u$.
4. The maps F_l and F_u are non-decreasing maps of degree one.

Evidently, a non-decreasing map of the real line of degree one may have some "flat spots" ("plateaus"). More precisely, for a degree one function $G : \mathbb{R} \to \mathbb{R}$ define the set Const(G) as the maximal open set on which G is a constant. We are especially interested in the G-orbits of points that avoid Const(G), and indeed one can show that such orbits exist.

Arguably, the modern theory of dynamical systems was founded by Poincaré in his papers [25] devoted to circle homeomorphisms. A beautiful concept of the rotation number of a circle homeomorphism introduced there can actually be introduced in our setting too. Before we state the corresponding result, let us introduce the concept of the *lifted cycle*.

Definition 3.1 A finite segment $Q = \{x, F(x), \ldots, F^{n-1}(x)\}$ of the orbit of a point $x \in \mathbb{R}$ under a map F is said to be a *lifted cycle* if $F^n(x) = x + k$ where k is an integer and the differences $F(x) - x, \ldots, F^{n-1}(x) - x$ are all non-integers. The number n is then called the *period* of Q.

Since $F(x + r) = F(x) + r$ for any integer r, then $F^n(F(x)) = F(F^n(x)) = F(x + k) = F(x) + k$; similarly, for any point y from the F-orbit of x we have $F^n(y) = y + k$. Moreover, if $z = x + r$ with r integer, then $F^n(z) = F^n(x) + r = x + k + r = z + k$. These observations justify the following notion.

Definition 3.2 In the situation of Definition 3.1, the union of all shifts of Q by multiples of k in both positive and negative directions is called the *lifted periodic orbit of x* (clearly, it contains the F-orbit of x). Moreover, the number $k/n = \rho_F(x)$ is called the *rotation number* and the pair (k, n) is called the *rotation pair* of Q (or x, or any point of the orbit of x, or any integer shift of any point of the orbit of x).

I turns out that rotation numbers can be defined for *all* (not only periodic) points of the real line under a non-decreasing degree one map.

Lemma 3.1 *If $G : \mathbb{R} \to \mathbb{R}$ is a continuous non-decreasing degree one map then for every $\in \mathbb{R}$ the limit*

$$\lim_{n \to \infty} \frac{G^n(x) - x}{n}$$

exists and is independent of x. This limit is a rational number if and only if G has a lifted cycle; moreover, in that case, the lifted cycle can be chosen so that the corresponding lifted periodic orbit is disjoint from $\mathrm{Const}(G)$. If the limit is p/q where p, q are coprime then for any lifted cycle Q of F the rotation pair of Q is (p, q).

We can now give the following definition.

Definition 3.3 Define the *rotation number* $\rho(F)$ of a non-decreasing map $F : \mathbb{R} \to \mathbb{R}$ of degree one as

$$\rho(F) = \lim_{n \to \infty} \frac{G^n(x) - x}{n}$$

where $x \in \mathbb{R}$ is any point.

The function $\rho : F \to \rho(F)$ acts nicely on the space of all non-decreasing continuous map of the real line of degree one.

Lemma 3.2 *The following properties hold.*

1. *The function $\rho : F \mapsto \rho(F)$ on the space of all non-decreasing continuous degree one maps of the real line is continuous.*
2. *If $H \leq G$ where at least one of the maps is non-decreasing then $H^n \leq G^n$ for every integer $n \geq 0$. If both H and G are non-decreasing, then $\rho(H) \leq \rho(G)$.*
3. *If $x \in \mathbb{R}$ is H-periodic and G is non-decreasing then $\rho_H(x) \leq \rho(G)$.*

4. *If $y \in \mathbb{R}$ is G-periodic and H is non-decreasing then $\rho(H) \le \rho_G(y)$.*

Suppose that a degree one map F of the real line is given. We have already defined the upper and the lower maps F_u and F_l for it (recall that these maps are non-decreasing). In fact we can define a continuous one-parameter family of non-decreasing degree one maps of the real line that slowly grow from F_l to F_u. Heuristically we can think of pouring water in all the "holes" of the graph of F from both top and bottom but not necessarily completely filling them up on either side and in this way gradually transforming F_l into F_u. Formal definition of the desired family of map follows.

Definition 3.4 Let $F : \mathbb{R} \to \mathbb{R}$ be degree one continuous map. Then we define the following one-parameter family of maps:

$$F_\mu = (\min(F, F_l + \mu))_u$$

where $0 \le \mu \le \mu_1 = \sup_{x \in \mathbb{R}} (F - F_l)(x)$.

The properties of maps F_μ are listed below.

Lemma 3.3 *Let F be a degree one continuous map of the real line. Then the family $\{F_\mu\}, 0 \le \mu \le \mu_1$ has the following properties.*

1. *Each map F_μ is continuous, of degree one, and non-decreasing.*
2. *$F_0 = F_l$ and $F_{\mu_1} = F_u$.*
3. *If $0 \le \lambda \le \mu \le \mu_1$, then $F_\lambda \le F_\mu$.*
4. *$F_\mu = F$ outside $\mathrm{Const}(F_\mu)$ and $\mathrm{Const}(F) \subset \mathrm{Const}(F_\mu)$.*
5. *Maps F_μ depend continuously upon μ.*

Let us now consider a degree one continuous map F of the real line to itself. We are interested in rotation numbers and rotation pairs of periodic points of F. Recall that $F_l \le F \le F_u$. Hence by Lemma 3.2 for every lifted periodic point x of F we have that $\rho(F_l) = l(F) \le \rho_F(y) \le \rho(F_u) = u(F)$. Therefore, we are guaranteed that rotation numbers of all periodic points of F belong to the interval $[l(f), u(f)]$ called the *rotation interval* of F and denoted by I_F. Next, let us investigate if *all* rational numbers in I_F are assumed as rotation numbers of a periodic point.

By Lemma 3.2, we have that $l(F) \le \rho(F_\mu) \le u(F)$. Moreover, by the same lemma the function $\psi(\mu) = \rho(F_\mu) : [0, \mu_1] \to I_F$ is continuous and increasing. Hence for each rational number $p/q \in I_F$, p, q coprime, there exists $\mu \in [0, \mu_1]$ with $\psi(\mu) = \rho(F_\mu) = p/q$. Since F_μ is continuous and non-decreasing (see Lemma 3.3), then by Lemma 3.1 the map F_μ has a lifted periodic orbit X of rotation number p/q and rotation pair (p, q). By Lemma 3.1, X can be chosen to avoid the set $\mathrm{Const}(F_\mu)$. By Lemma 3.3 this implies that in fact X avoids $\mathrm{Const}(F)$ and therefore X is a lifted periodic orbit of F itself. Hence indeed every rotation pair (p, q) with coprime p, q such that $p/q \in I_F$ is realized on a lifted periodic orbit of F.

Now, if we are given a circle map $f : S^1 \to S^1$ of degree one, we can lift it to a degree one map $F : \mathbb{R} \to \mathbb{R}$ of the real line (the lifting is not unique as we explained above, but we can fix one such lifting). Since maps F_μ are non-decreasing, the lifted periodic orbits of F_μ's discussed above, when put back onto the circle, become periodic orbits of f which combinatorially behave like rigid rotations (i.e., can be conjugated into rigid rotations by orientation preserving circle homeomorphisms). Together with some important additional arguments all this yields the full description of the periods of circle maps of degree one which follows below.

Definition 3.5 Suppose that $a, b \in \mathbb{R}$ are two real numbers; also, suppose that $l \in \mathbb{Z} \cup 2^\infty$. Then set $M(a, b) = \{n \in \mathbb{Z}^+ : \text{there exists } k \in \mathbb{Z} \text{ such that } a < k/n < b\}$. Moreover, define the set $S(a, l)$ as follows: If a is irrational then $S(a, l)$ is empty, and if $a = k/n$ where k and n are coprime then $S(a, l) = \{ns : s \in \text{Sh}(l)\}$.

The main result of [23] is the following theorem.

Theorem 3.2 ([23]) *Suppose that $f : S^1 \to S^1$ is a continuous map of degree one. Then there exist real numbers $a \le b$ and $l, r \in \mathbb{Z}^+ \cup \{2^\infty\}$ such that the set of periods of f coincides with the set $M(A, b) \cup S(a, l) \cup S(a, l)$. Moreover, for every set A in the above form, there exists a continuous circle map of degree one such that its set of periods coincides with A.*

To illustrate the meaning of this theorem denote by $i(x)$ the integer part of a real number x and by $\{x\} = x - i(x)$ the fractional part of x. Consider the following map of the real line of degree 1:

$$F(x) = \begin{cases} 2x \, (0 \le x \le 2/3), \\ 2 - x \, (2/3 \le x \le 1), \\ i(x) + F(\{x\}) \, (x \notin [0, 1]) \end{cases}$$

Let us show that $I_F = [0, 1/2]$. Indeed, it is easy to see that the following holds:

$$F_l(x) = \begin{cases} 2x \, (0 \le x \le 1/2), \\ 1 \, (2/3 \le x \le 1), \\ i(x) + F(\{x\}) \, (x \notin [0, 1]) \end{cases}$$

and

$$F_u(x) = \begin{cases} 1/3 \, (0 \le x \le 1/6), \\ 2x \, (1/6 \le x \le 1/2), \\ 1 \, (1/2 \le x \le 1), \\ i(x) + F(\{x\}) \, (x \notin [0, 1]) \end{cases}$$

which allows us to see that (a) since $F_l(0) = 0$ then $l(F) = 0$ and (b) since $F_u(1/3) = 2/3$, $F_u(2/3) = 4/3 = 1/3 + 1$ then $1/3 \mapsto 2/3 \mapsto 4/3$ is a lifted periodic orbit of F_u and hence $u(F) = 1/2$. Thus, the rotation interval of F is $[l(F), u(F)] = [0, 1/2]$, and, moreover, periods 1 and 2 are among periods of periodic points of F. Let us now consider any positive integer $n > 2$. Then, since $0 < 1/n < 1/2$ it follows that $n \in M(0, 1/2)$ and hence, by Theorem 3.2, there exists an F-periodic point of period n. We conclude that F has lifted periodic orbits of all periods.

The corresponding degree one map of the unit circle $f : S^1 \to S^1$ can be described as follows. Let us normalize the unit circle so that its length is 1; also, speaking of points of S^1 we will interchangeably talk about angles. Define f on the arc $K = [0, 2/3]$ so that 0 is an f-fixed point and otherwise all angles in K are doubled. Then K expands twofold and covers the entire S^1 (in fact, the arc $[0, 1/3]$ is covered by K twice). Now, on the remaining arc $L = [2/3, 1]$ our map f acts as a mere "flip", i.e. without any expansion or shrinking it simply reflects L with respect to the angle 0 so that its f-image is the arc $[0, 1/3]$. By the previous paragraph, it follows that f has periodic orbits of all periods.

3.2 Maps of the n-od

After the Forcing Sh-Theorem, it is natural to ask what other topological spaces admit concise descriptions of sets of periods of periodic points of their continuous self-mappings. Evidently, higher dimensional spaces, or non-compact spaces, will have very different properties, too different to expect similar or even comparable in terms of their completeness results. Therefore, a lot of efforts were concentrated upon one-dimensional topological spaces. Arguably, after the closed interval, the simplest ones here are the circle and the n-od (the latter can be described as the union of n closed intervals called *branches* that all have a common endpoint v called the *vertex* and are otherwise pairwise disjoint). In this section, we consider the results giving a full description of periodic points of continuous self-mappings of the n-od.

Observe that the closed interval can be viewed as the 1-od, or, better yet, as the 2-od. Moreover, consider a continuous map f of the interval $I = [0, 1]$ to itself. If it has no fixed points in $(0, 1)$ then, clearly, all points of the open interval $(0, 1)$ are moving in the same direction. Thus, the map has either one periodic point or two periodic points, and all of its periodic points are fixed. Removing this trivial case, we may assume that f has a fixed point $a \in (0, 1)$. Then we can think of I as the 2-od with vertex a and two branches $[0, a]$ and $[a, 1]$. Thus, as a direct extension of the interval maps, we can consider the maps of the n-od that fix its vertex.

Historically, the first results in this direction were obtained in [3] where maps of the 3-od were studied. A little later these results were generalized in [4] where the problem was completely solved and the full description of sets of periods of periodic points of continuous self-mappings of the n-od was given. In our presentation, we follow [4], thus for brevity references to [4] are mostly omitted in the rest of this section.

In the Forcing Sh-Theorem, the forcing relation among periods of periodic points of continuous interval is established. However, the forcing relation among periods is just a tool (though important) leading to the ultimate result, the complete description of sets of periods of interval maps. Already in case of the circle maps of degree one we saw that, strictly speaking, there is no universal order among integers that would describe forcing relation (in that case one introduces rotation numbers of periodic orbits and shows that rotation numbers located between given rotation numbers are themselves rotation numbers of periodic points). Thus, the description of the sets of periods can be obtained in other ways, not necessarily through forcing relation. The maps of the n-od give a similar example: Even in the absence of the usual forcing relation, one can fully describe sets of periods in that case.

Definition 3.6 Define a series of partial orderings $\leq_n$ among the natural numbers. The partial ordering $\leq_1$ coincides with the Sh-ordering: $m \leq_1 n$ if only if $n \succ m$. Assume that $s > 1$ is an integer. Then one defines $\leq_s$ as follows.

1. We have $m \leq_s 1$ if and only if $m = 1$.
2. Suppose that $k = ts$ where t is an integer (in other words, k is divisible by s). Then $m \leq_s k$ if and only if either $m = 1$ or $m = rs$ is divisible by s too and $r = m/s \leq_1 t = k/s$.
3. Suppose that $k > 1$ is not divisible by s. Then $m \leq_s k$ if and only if $m = 1$, $m = k$, or $m = ik + js$ for some integers $i \geq 0$, $j \geq 1$.

Definition 3.6 shows that, in the beginning of $\leq_s$, we have 1, and then all the multiples of s ordered according to the Sh-order. They are ordered this way because of parts 1. and 2. of Definition 3.6; the fact that any number $k > 1$ not divisible by s and any number $m = st$ are such that $m \leq_s k$ follows because $m = 0 \cdot s + st$. It remains to understand the $\leq_s$-order among the numbers greater than 1 that are not multiples of s. These non-multiples of s are arranged as $n - 1$ chains according to their residues modulo s, and in each chain the $\leq_s$-order is opposite the usual order. Indeed, by part 3. of Definition 3.6 we have that $m + s \leq_s m$ which implies the claim. Finally, part 3. of Definition 3.6 also implies other relations among elements of different residue classes modulo s.

Definition 3.7 A set E of positive integers is an *initial segment* of $\leq_s$ if whenever $k \in E$ and $m \leq_s k$, $m \in E$.

We are now ready to state the main result of [4].

Theorem 3.3 *Let f be a continuous self-mapping of the n-nod. Then the set of periods of periodic points of f is a non-empty finite union of initial segments of partial orders $\{\leq_s : 1 \leq s \leq n\}$. Conversely, if E is a non-empty finite union of initial segments of partial orders $\{\leq_s : 1 \leq s \leq n\}$, then there is a continuous self-mapping of the n-od which fixes the vertex and is such that the set of periods of its periodic points is E.*

The rest of this section is devoted to the sketch of the proof of the first claim of Theorem 3.3. Moreover, from now we will make the standing assumption that the vertex of the n-od is fixed. We emphasize here that we do this solely for the sake of brevity and simplicity; Theorem 3.3 holds for all continuous self-mappings of the n-od without any assumptions. However, first we want to state another result from [4] (again, we state it under our standing assumption while in [4] it is obtained without any assumptions at all). From now on let us denote the n-od by X_n and its vertex by v; we are considering maps $f : X_n \to X_n$ such that $f(v) = v$. Also, we will denote branches of X_n as capital B with a subscript. In the definitions below, we use notation in which the dependence upon f is omitted for the sake of brevity.

Definition 3.8 Suppose that f has a periodic orbit $A \neq \{v\}$. Choose the branches $B_1, \ldots, B_k$ that are non-disjoint from A. Let z_i be the closest to v point of A on B_i. Define a map $\psi : \{1, \ldots, k\} \to \{1, \ldots, k\}$ so that $f(z_i) \in B_{\psi(i)}$. Then ψ must have one or more periodic orbits. If ψ has a periodic orbit (of branches) of period p we say that A is of *type* p.

Observe that a periodic orbit can have several types (depending on how many periodic orbits the map ψ has).

Theorem 3.4 *If a map f has a periodic point of period k and type p, then for every $m \leq_p k$ the map f has a periodic point of period m.*

If A is of period k then $p \leq k$ for any type p of A. Hence $\leq_p$ in Theorem 3.4 only applies to integers greater than $p - 1$. We will now prove Theorem 3.4 fixing a periodic orbit A of type p. First we need to introduce the convenient notation.

Definition 3.9 (Notation) Let $x_0 = v$, $A = \{x_1, \ldots, x_k\}$, and $A' = \{x_0, \ldots, x_k\}$. Let $B_1, \ldots, B_n$ be branches of X_n. We will call B_i *trivial* if it is disjoint from A and non-trivial otherwise. Since A is of type p, we may assume that for every j, $1 \leq j \leq p$, $x_j \in B_j$ is such that $[v, x_j] = I_j$ contains no points of A in (v, x_j), $f(x_j) \in B_{j+1}$ $(1 \leq j \leq p - 1)$, and $f(x_p) \in B_1$. Thus, $B_1, \ldots, B_p$ are non-trivial branches. A *basic* interval is an interval $[x_i, x_j]$ where $x_i, x_j \in A'$ and the interval (x_i, x_j) is disjoint from A' (thus, each I_j is basic). The remaining k basic intervals are numbered arbitrarily as $I_{p+1}, \ldots, I_k$.

Now we are ready to define a directed graph generated by A.

Definition 3.10 (Directed graph of A) Construct the graph G as follows. Its vertices are the basic intervals $I_1, \ldots, I_k$. If $I_r = [x_s, x_t]$ then $I_r \to I_l$ is an arrow in G if and only if $I_l \subset [f(x_s), f(x_t)]$. A *loop* of length m is a sequence $J_0 \to \cdots \to J_m = J_0$ in G. It is *non-repetitive* if it cannot be written as a single smaller loop repeated more than once.

The graph G does not reflect all the dynamics of f as in reality the f-image of a basic interval $I_r = [x_s, x_t]$ may cover more than just $[f(x_s), f(x_t)]$. Thus G reflects the minimal guaranteed coverage of basic intervals by basic intervals under f. The map g defined below is in more direct connection with G than f.

Definition 3.11 (Linear maps) Let $g : X_n \to X_n$ be the map coinciding with f on A', linear on each basic interval, and constant on each component of the complement of the union of the basic intervals.

Similarly to how it is done in Chap. 1, one can establish a connection between non-repetitive loops in G and periodic points of f.

Lemma 3.4 *If $J_0 \to \cdots \to J_m = J_0$ is a non-repetitive loop in G with at least one interval not containing v then f has a point $x \in J_0$ of period m such that $f^i(x) \in J_i$, $0 \leq i \leq m - 1$. On the other hand, if g has a point of period m with $m \neq 1$, $m \neq k$ then there is a non-repetitive loop of length m in G. In particular, if g has a point of period m, then so does f.*

A point from the first claim of Lemma 3.4 is said to *follow* the corresponding loop. By Lemma 3.4 we can now consider the piecewise-linear map g rather than f and study its periodic points.

Proof of Theorem 3.4. Clearly, G admits an important loop $a = \{I_1 \to I_2 \to \cdots \to I_p \to I_1\}$. In addition to that, construct a loop b in G of length k as follows. Start with the basic interval $I_p = [v, x_p]$ and consider a string of intervals $[v, g^i(x_p)]$ until we get back to $[v, g^k(x_p)] = [v, x_p]$. Then we can find a basic interval $J_{k-1} \subset [v, g^{k-1}(x_p)]$ such that $J_{k-1} \to [v, x_p]$ is an arrow in G. Using induction, we can step by step find a loop $b = \{[v, x_p] = J_0 \to J_1 \to \cdots \to J_{k-1} \to J_0\}$ of length k. Divide the rest of the proof into cases.

(1) Suppose that k and m are not multiples of p. Assume that $m \le_p k$ and show that f has a point of period m. Then by definition $m = ip + jk, i \ge 1, j \ge 0$. Since m is not a multiple of p, then $j \ge 1$. Then the loop $c = a^i b^j$ has length m. Let us show that b contains a basic interval that does not contain v. Indeed, otherwise b has to be a repetition of a, a contradiction with the assumption that k is not a multiple of p. Now, if c is non-repetitive, then by Lemma 3.4 we get a point of period m from the loop c.

Suppose now that c is repetitive. If the repeated subloop of c is contained in a^i then it is a repetition of several copies of a. Hence the entire c is a repetition of several copies of a and m is a multiple of p, a contradiction. Hence the repeated loop in c is of the form $a^i d$ for some d and the remaining b^j can be represented as $da^i e$. Therefore we can replace c by $a^{2i} de = c'$ and continue in the same fashion, concentrating copies of a in the beginning of the loop. As we saw above, a repetitiveness of a loop implies that in its second part there are copies of a, and we can repeat the argument. Thus in the end we will get a non-repetitive loop and will then find a point of period m.

(2) Suppose that $g(B)$ is contained in a branch of X_n for each non-trivial branch B of X_n. Assume that $m \le_p k$ and show that f has a point of period m. Indeed, k is a multiple of p and non-trivial branches of X_n are cyclically permuted under f. It is easy to see then that the Forcing Sh-Theorem and our definition of the order $\le_p$ imply the existence of a point of period m.

(3) Suppose that there is a non-trivial branch B such that the smallest connected subset $T \subset X_n$ of X_n, containing $g(B)$, contains v. We claim that then g has points of every multiple m of p. Indeed, it is easy to see that $g^{p+j}(0, x_1) \supset g^j(0, x_1]$ for every j. On the other hand, there must exist i such that $g^i(0, x_1]$ contains 0. Indeed, otherwise the branches B_r map completely into the branch B_{r+1} because this is where $g(x_r)$ is, and, similarly, $g(B_p) \subset B_1$. However, this contradicts the assumptions of case (3). Fix the least i with the above stated property, and consider subcases.

(3.a) Suppose that $i \ge p$. Set $K = g^{i-p}[v, x_1]$ and $J = g^p[v, x_1]$. Clearly, $g^p(K) = J$. Let us show that there exists $y \in K$ such that $g^p[v, y] = K$. First, assume that $i = p$ and $K = [v, x_1]$. Then the claim easily follows from our assumptions on dynamics of branches $B_1, \ldots, B_p$. Now, suppose that $i > p$ and set $K = [v, w]$. Then $i - p > 0$ and we can consider $T = g^{i-p-1}[v, x_1]$; clearly, v is an endpoint of T. Moreover, $g(T) = K$ and $g^{p-1}(K) = S = [v, x_s] \supset T$ for some x_s

(the latter follows from the minimality of i). Since $g(v) = v$, then we can find $y \in K$ such that $[v, y]$ has the property $g^{p-1}[v, y] = T$. and hence $g^p[v, y] = K$. If we choose the smallest segment $[v, y]$ with this property, we will see that $f^p(y) = w$.

On the other hand, the choices we made imply that there exists the closest to y point $z \in [y, w]$ such that $g^p(z) = v$. Hence, the closed intervals $[v, y]$ and $[y, w]$ form a *horseshoe* for g^p (their interiors do not intersect while their g^p-images contain them both) and, moreover, their images are contained in $B_1, \ldots, B_{p-1}$ until the g^p-image equal to K for both. It follows that g (and, hence, f) has periodic points of all periods that are multiples of p.

(3.b) Let $i < p$; by definition this means that intervals $(v, x_1]$, $g(v, x_1]$, $\ldots$, $g^{i-1}(v, x-1]$ do not contain v while $v \in g^i(v, x_1] = J$. Choose a subinterval K of $[0, x_i]$ with $g^{p-i}(K) = [0, x_1]$; then, clearly, $K \subset J$ and $g^p(K) = J$. Repeating the arguments from (3.a) shows that in this case, too, g has points of all periods that are multiples of p.

It now immediately follows from how $\leq_p$ is defined that if $m \leq_p k$ then g (and, hence, f) has a periodic point of period m. Thus, given a map $f : X_n \to X_n$ one can consider all types p of all its periodic points, construct the corresponding initial segments of $\leq_p$-orderings, and in this prove the first part of Theorem 3.3. We omit the other direction of the theorem in which self-mappings of the n-od with given sets of periods of their periodic points are constructed.

3.3 Other Graph Maps

For the sake of brevity, and slightly abusing the language, in the rest of this section, we will use the following terminology.

Definition 3.12 A *graph* is a compact connected topological space G that is the union of finitely many arcs whose pairwise intersections consist of either one or two points. A *tree* is a graph that contains no Jordan curves. For every point x of a graph G there exists a small neighborhood U of x in G such that all neighborhoods $V \subset U$ of x are homeomorphic to the same n-od. Then the number n is said to be the *order* (*of x in G*). If $n \geq 3$ then the point x is said to be a *vertex* of G; if $n = 1$ then x is said to be an *endpoint* of G. The number of endpoints of a graph G is denoted by $e(G)$; the number of vertices of G is denoted by $v(G)$.

Evidently, the circle is a graph while the interval and the n-od are trees. Slightly more appropriate term for graphs would be *one-dimensional compact branched manifolds*; we choose our terminology for the sake of brevity.

The most challenging problem in the field is to give a full description of possible sets of periods of periodic points of continuous self-mappings of a tree/graph in terms of the tree/graph in question. This problem has not been solved yet in full even though some important results were obtained. Below we describe the most interesting (in our view) such results.

We would like to emphasize that our choice of articles referenced in this section is quite selective. There are, indeed, a lot of interesting and deep results describing periodic behavior of one-dimensional maps. Even touching upon all of them, let alone covering them in detail, would require a book of much greater length. Therefore, we admit, there is no pretension of completeness in how we cover the material. Our selection of the articles surveyed reflects our desire to include the results which, in our view, answer certain questions in full and leave the reader with the feeling of closure. Reading about them will help the reader to paint a full picture of at least some aspects of the entire topic of periodicity of points in one-dimensional dynamics. Also, we tried to stay close to the historical development of the subject. Hopefully, this explains our choice of the material to the reader.

3.3.1 Graph-Realizable Sets of Periods

In the setting mentioned above, one considers a given graph and then aims at describing periods of continuous maps on that graph. However, it is also interesting to see what possible sets of periods can appear in general, without specifying the graph. Can any set of integers be the set of periods of a continuous self-mapping of some graph? This natural question leads to the following definition.

Definition 3.13 ([14, 16]) A set $A \subset \mathbb{N}$ is a *graph-realizable set (of periods)* if and only there exist a graph X and a continuous map $f : X \to X$ such that the set of periods of periodic points of the map $f : X \to X$ is A.

It turns out that because of structural properties of continuous self-mappings of graphs, the family of graph-realizable sets of periods is far smaller than, say, the family of all possible sets of natural numbers. It was originally described in the preprint [14] and later published in [16]. To state the results of [14, 16], we need a few more definitions.

Definition 3.14 ([14, 16]) Set $lZ \equiv \{li : i \geq 1\}$, $Q(n) \equiv \{2^i n : i \geq 0\}$. Also, say that a set A *almost coincides* with a set B if the symmetric difference $(A \setminus B) \cup (B \setminus A)$ is finite.

The main result of [14, 16] describes graph-realizable sets.

Theorem 3.5 ([14, 16]) *A set $A \subset \mathbb{N}$ is graph-realizable if and only if it is finite or coincides with a finite union of some sets lZ or $Q(n)$.*

The proof here is based on earlier structural results on dynamics on graphs [10–13], see also [15]. We summarize them below. If $f : G \to G$ is continuous and G is a graph, the closure of the set of its periodic points can be represented as the union of sets of three types. This representation is called the *spectral decomposition*.

First, there are the *basic* sets. Each basic set is generated by a *cycle of subgraphs*, that is the finite union $A = \bigcup_{i=1}^{n} G_i$ of subgraphs with at most finite pairwise intersections cyclically permuted by the map. Let us call such a cycle of subgraphs *basic*.

The number n is then said to be the *period* of A. The basic set $B(A)$ generated by A could be of the following two types.

(1) It could coincide with A in which case $f|_A$ is transitive while $f^n|_{G_i}$ is mixing for $1 \leq i \leq n$. Moreover, $f^n|_{G_i}$ then has the so-called *specification* property introduced by R. Bowen in [17].

(2) Otherwise $B(A)$ is a Cantor set consisting of all points of A whose arbitrary neighborhoods in A have dense orbits in A. In that case the map on A is monotonically semiconjugate to a transitive map g on a cycle of subgraphs and the properties listed in the previous paragraph apply to this transitive map. In particular, the appropriate power of g that fixes each subgraph in the cycle, has specification property. Closures of components of $A \setminus B(A)$ can be wandering, preperiodic, or periodic. The restriction of f on a cycle of closures of components of $A \setminus B(A)$ is a map to which one can apply similar arguments to find other basic sets inside this cycle, and so on.

Since the appearance of [17] dynamical systems with the specification property have been studied in great detail and a lot of far reaching corollaries for them have been proven. These are related, in particular, to the structure of the space of invariant measures of maps with specification (see, e.g., [19] and references therein). In the end, these results, combined with the spectral decomposition, imply, e.g., the following. Any graph self-mapping may have finitely many invariant sets on which the map is monotonically semiconjugate to a permutation of circles with appropriate power of the map being an irrational rotation. These sets are said to be *circle-like*. It turns out that for any graph map f invariant measures concentrated on a periodic orbit, together with finitely many invariant measures concentrated on circle-like invariant sets, form a dense subset of all ergodic measures. Since, in the tree case, there are no circle-like invariant sets, for any tree map f invariant measures concentrated on one periodic orbit, are dense in the space of all invariant ergodic measures of f.

For us, though, an important corollary of the specification property is that maps with specification have periodic points of all but finitely many periods. This implies that *the set of periods of periodic points of $f|_A$ almost coincides with the set $n\mathbb{Z}$.*

There are also the *solenoidal* sets, otherwise called *generalized adding machines*. Consider a nested family of cycles of subgraphs of growing to infinity periods; the intersection of these cycles of subgraphs is a *solenoidal set* while the cycles of subgraphs are said to be *generating*. Some generating cycles of subgraphs may be contained in basic cycles of subgraphs. Then the periods of the periodic points contained in them can add at most finitely many periods to the entire set of periods, and can therefore be ignored. The only possibly existing solenoidal sets that are not contained in basic cycles of subgraphs must be such that the periods of their generating subgraphs form a sequence almost coinciding with a sequence $Q(n)$ for some n. Finally, there are also some periodic orbits that are not contained in basic cycles of subgraphs; however, it is proven in [16] that they exhibit at most finitely many periods and periods forming sets $Q(s)$ for finitely many values of s. Putting together these facts, ones get one direction of Theorem 3.5. The other direction is proven by a special construction.

Observe that the sets of periods of self-mappings of the interval, maps of degree one of the circle to itself, and continuous maps of the n-od, are of the types described in Theorem 3.5. Indeed, in the interval case, the set $P(f)$ of periods of an interval map f can be $\text{Sh}(k)$ where k is an integer or $k = 2^\infty$. If k is odd, then $P(f)$ almost coincides with the set of all natural numbers $1Z$; if $k = (2l + 1)2^n$ where $l \geq 1$ then $P(f)$ almost coincides with $2^n Z$; if $k = 2^\infty$ then $P(f)$ coincides with $Q(1)$; finally, if $k = 2^m$ then $P(f)$ is finite.

In the case of a circle self-mapping g of degree one, the set $P(g)$ of periods of periodic points of g depends on the rotation interval $[u, v]$ of g. Consider first the case when $[u, v]$ is non-degenerate. Then, evidently, for all large enough N there will be rational numbers $M/N \in (u, v)$ which implies that $N \in P(g)$. In other words, in this case, $P(g)$ almost coincides with the set $1Z$ of all natural numbers. If $u = v = p/q$ is rational then $P(g) = q \cdot \text{Sh}(k)$ is the set of products of q and all integers from the set $\text{Sh}(k)$ for some k. By the previous paragraph $P(g)$ is of necessary form. Finally, if $u = v$ is irrational then $P(g)$ is empty.

In the case of a continuous self-mapping h of the n-od the set of periods $P(h)$ of h is a finite non-empty union of initial segments of orders $\leq_p$ for various $p \leq n$. Hence it is sufficient to show that given $p \leq n$ any initial segment of $\leq_p$-order is of the form described in Theorem 3.5. Choose $k \geq 1$ and consider the initial $\leq_p$-segment A consisting of all numbers m with $m \leq_p k$. Then definition and because of our analysis of the sets $\text{Sh}(i)$ we see that the desired follows if k is a multiple of p. Suppose that k is not a multiple of p. Then A includes 1 and all sets of the form $ik + jp$ for some integers $i \geq 0, j \geq 1$. It follows that the set of these numbers $ik + jp$ almost coincides with the set qZ where q is the greatest common divisor of k and p. Thus, in this case, the set $P(h)$ is of desired form.

3.3.2 Trees

Let us now consider tree maps. In what follows given a tree T let us denote by $v(T)$ the number of its vertices and by $e(T)$ the number of its endpoints. E.g., for the n-od X_n, $n \geq 3$ we have that $v(X_n) = 1$ while $e(X_n) = n$; for the interval I we have $v(I) = 0$ while $e(I) = 2$.

The results concerning sets of periods for trees and their self-mappings are more specific than the results from the previous subsection as they take into account the structure of the tree and describe, based upon this structure, possible sets of periods. We begin with the paper [6] where the results of [4] are generalized for continuous self-mappings of trees *with all vertices fixed*. Unlike in the case of the n-od, in case of an arbitrary tree the fact that vertices are fixed is essential. To illustrate it, the following example is suggested in [5]. Consider a tree H whose shape is like the shape of the letter H; in particular, we have $v(H) = 2$ and $e(H) = 4$. Visualize H as, indeed, letter H in its usual position.

Denote its vertices by l and r (from "left" and "right"); denote its endpoints by a_l ("above left"), b_l ("below left"), a_r ("below right"). Then put two points on the horizontal plank of H and denote them x and y so that $l < x < y < r$. Define the map $f : H \rightarrow H$ as follows:

$$f(l) = f, f(r) = l; f(x) = y, f(y) = a_l, f(a_l) = a_r, f(a_r) = b_l, f(b_l) = b_r, f(b_r) = x$$

Clearly, f has a fixed point, a periodic orbit A of period two ($A = \{l \rightarrow r \rightarrow l\}$), and a periodic orbit B of period 6 ($B = \{x \mapsto y \mapsto a_l \mapsto a_r \rightarrow b_l \rightarrow b_r \rightarrow x\}$). Define the map f on the entire H so that every component of $H \setminus (A \cup B)$ maps forward in a linear fashion. Clearly, there exists a fixed point $z \in [x, y]$, and z is the only fixed point in $[x, y]$.

Let us study the dynamics of f. The following shows how all components of $H \setminus (A \cup B)$ except for $[x, y]$ map forward under f:

$$[x, l] \mapsto [y, r] \mapsto [a_l, l] \mapsto [a_r, r] \mapsto [b_l, l] \mapsto [b_r, r] \mapsto [x, l]$$

which means that the entire segment $[x, l]$ maps onto itself as the identity map. Clearly, all its points are of period 6 except for l which is of period 2. Otherwise, there is a segment $[x, y]$ that expands under f and whose all points eventually leave it except for the fixed point z. Thus, indeed, f has only periodic points of periods 1, 2 and 6. However, it will follow from the main theorem of [6] that no continuous map $g : H \rightarrow H$ that fixes vertices of H can have the set of periods of its periodic orbits $\{1, 2, 6\}$.

Theorem 3.6 ([6]) *Let T be a tree. Then the following holds.*

1. *Let $f : T \rightarrow T$ be continuous, with all vertices of T fixed. Then the set of periods of all periodic points of g is a non-empty union of initial segments of $\{\leq_p : 1 \leq p \leq e(T)\}$.*
2. *Conversely, if S is a non-empty finite union of initial segments of $\{\leq_p : 1 \leq p \leq e(T)\}$ then there is a continuous map $f : T \rightarrow T$ with all the vertices fixed such that its set of periods of its periodic orbits is S.*

In the proof of Theorem 3.6, the authors first prove that if $f : T \rightarrow T$ is a continuous tree map that has a periodic orbit P of period k, then there exists p, $1 \leq p \leq e(T)$ such that for all m, $m \leq_p$ the map f has a periodic orbit of period m. To do this, they first consider the smallest subtree $[P]$ of T containing P. Using natural retraction, one can reduce the problem to the situation when $T = [P]$. Then the authors introduce tools that have been used in [4] for maps of the n-od, such as basic intervals of loops of basic intervals. Also, they introduce a metric on T that is additive on intervals, define linear maps in the sense of this metric, and show that it is sufficient to consider only such linear maps.

The proof of claim (1) of Theorem 3.6 then proceeds by induction on the number of endpoints and on the number of vertices of a tree. By the Forcing Sh-Theorem and by the results of [4] it follows that Theorem 3.6 holds for the interval and for the n-od for every n. By induction, the authors assume that Theorem 3.6 holds for all trees T' such that $e(T') \leq e$, $v(T') \leq v$, and at least one of these inequalities is strict. Then they prove that Theorem 3.6 holds for any tree T with $e(T) = e$ and $v(T) = v$.

Let us fix T, points e, v, periodic orbit $P \subset T$ of period k, and a continuous map $f : T \to T$. The arguments begin by taking care of some trivialities. Thus, we may assume that f has no subtrees on which f is the identity. Moreover, we may assume that every interval in T connecting two vertices of G contains a point of P. Indeed, otherwise the map on this interval is the identity (recall that we consider continuous maps that fix all vertices), and everything follows from the above. Thus, a clever usage of induction and the assumptions about the maps allows to reduce an arbitrary map to a map from much easier to deal with class. In the end, careful considering of loops of basic intervals similar to [4] allows to complete the proof of Theorem 3.6.

Let us comment that the condition of the map f fixing all the endpoint of the tree T can be replaced with a similar condition of f permuting the endpoints of T. For such tree maps a similar realizability result can be proved.

Theorem 3.7 ([7]) *Let T be a tree with n vertices. Let $f : T \to T$ be continuous and suppose that the n vertices form a periodic orbit under f. Then*
1. *a. If n is not a divisor of 2^k then f has a periodic point with period 2^k.*
    b. If $n = 2^p q$, where $q > 1$ is odd and $p \geq 0$, then f has a periodic point with period $2^p r$ for any $r \geq q$.
    c. The map f also has periodic orbits of any period m where m can be obtained from n by removing ones from the right of the binary expansion of n and changing them to zeroes.
2. *Conversely, given any n, there is a tree with n vertices and a vertex map f that has no other periods apart from the ones given above.*

The next paper we want to mention is [1]. In this paper, the authors consider arbitrary continuous maps of a given tree T and give a more precise description of the sets of periods of continuous maps $f : T \to T$. The description they provide is rather involved and complicated, thus we choose to interpret their main results, losing in precision but gaining in transparency. To describe the results of [1], we need a few new concepts.

Consider a tree T. Then, according to [1], for each continuous map $f : T \to T$ there exists a finite collection of finite sequences of numbers which depend on f and, through a certain construction which involves a few other parameters, define the set of periods of f. Fix one such sequence $\widetilde{S} = \{p_1, p_2, \ldots, p_m\}$ from the collection $\mathbf{S}$. Set $\Pi(\widetilde{S}) = p_1 \cdots p_m$. This number plays an important role in describing the set of period of f, however there are a few other numbers and sets of numbers too. The next set of numbers we need is $K(\widetilde{S}) = \{p_1, p_1 \cdot p_2, \ldots, p_1 \cdots p_{m-1}\}$. Also, there exists another finite set of numbers $F(\widetilde{S})$ with specific and complicated properties that restrict possibilities for $F(\widetilde{S})$ but do not determine exactly what it is. Finally, there is a number $\lambda_{\widetilde{S}}$ that defines the collection $\Pi(\widetilde{S})\{2, 3, \ldots, \lambda_{\widetilde{S}}\}$ of all multiples of $\Pi(\widetilde{S})$ from $2\Pi(\widetilde{S})$ through $\lambda_{\widetilde{S}} \cdot \Pi(\widetilde{S})$.

As we can see from the previous paragraph, finite sequences $\widetilde{S}$ are rather important. Each such sequence comes from a collection of finite sequences Σ_T which we will now describe. Namely, suppose that $(p_1, \ldots, p_m) \in \Sigma_T$. This means that there exists a sequence of subtrees $S_1 \supset S_2 \supset \cdots \supset S_m$ with the following properties.

(1) For each $i \leq m - 1$ the tree S_i contains a tree W_i with p_i endpoints such that one of components of $S_i \setminus \mathrm{Int}(W_i)$ coincides with S_{i+1} and each component of $S_i \setminus \mathrm{Int}(W_i)$ has at least $e(S_i)$ endpoints. Also, $e(S_m) \geq p_m$.

(2) The tree S_i, $i \leq m$, is not a k-od for any k.

It is shown in [1] that Σ_T is a finite collection of finite number sequences. Let us call them *admissible*.

Theorem 3.8 ([1]) *Suppose that $f : T \to T$ is a continuous map of a tree T to itself. Then there exists a finite collection $\mathbf{S} \subset \Sigma_T$ of finite admissible sequences such that the set of periods of periodic points of f can be expressed in the following form:*

$$\bigcup_{\widetilde{S} \subset \mathbf{S}} K_{\widetilde{S}} \cup F_{\widetilde{S}} \cup \mathscr{I}_{\widetilde{S}} \setminus \Pi(\widetilde{S})\{2, 3, \dots, \lambda_{\widetilde{(S)}}\}$$

where $\mathscr{I}_{\widetilde{S}}$ is an initial segment of $\leq_{\Pi(\widetilde{S})}$-ordering such that each $\leq_{\Pi(\widetilde{S})}$-maximal element of $\mathscr{I}_{\widetilde{S}}$ belongs to $\{1\} \cup (p_1 p_2 \cdots p_{m-1} Z \cup 2^\infty)$.

Theorem 3.8 can be viewed as one specifying results on graph-realizable sets of periods discussed in the previous subsection. The comparison of the two results shows advantages and drawbacks of them both. The results concerning graph-realizable sets are more abstract and establish the structure of the sets of periods of any self-mapping of any graph. An important drawback here is the fact that the connection between the properties of the graph and the properties of the set of periods are not connected with each other. Besides the set of periods is found up to a constant set. Still, these results are obtained for *any graph*. On the other hand, Theorem 3.8 allows for a much more precise, almost precise, description of sets of periods (even though in some cases this is still not a complete description because of an incomplete description of the finite sets $F_{\widetilde{S}}$) of *tree* maps.

3.3.3 Graphs With Exactly One Loop

Let G be a graph whose number of vertices $v(G)$ equals the number of edges $e(G)$. This is equivalent to saying that G has precisely one loop, or in more fancy language, that the *Euler characteristic* of G is zero. Examples of such a graphs include a circle, the letters σ, P, A etc. For such graphs, and for the maps $G \to G$ fixing branch points one can obtain the following characterization of possible sets of periods.

Theorem 3.9 ([21]) *Let G be a connected graph with $e(G) = v(G)$, and such that G is not a circle. Then*

1. *Let f be a continuous map with all branching points fixed. Then the set of periods of f is a nonempty union of initial segments of $\{\leq_s: 0 \leq s \leq e(G) + 2\}$.*
2. *Conversely, if S is a nonempty finite union of initial segments of $\{\leq_s: 0 \leq s \leq e(G) + 2\}$, then there is a continuous map $f : G \to G$ with all the branching points fixed such that the set of periods of f is precisely S.*

Here the partial orderings $\leq_s$ for $s \geq 1$ are defined as in the section on n-ods (see the Definition 3.6). The ordering $\leq_0$ is defined by $1 \leq_0 \ldots 5 \leq_0 4 \leq_0 3 \leq_0 2$, i.e., it is the converse of the usual ordering of the positive integers with the exception that 1 is the smallest number, rather than the greatest. As before, an initial segment of an ordering $\leq_s$ is a set containing all $m \leq_s k$, whenever it contains k.

3.3.4 Figure Eight Graph

When we move on with the zoo of graphs G and consider graphs with two (or more) cycles, the problem of description of all possible sets of periods of a map $G \to G$ becomes very hard. One of the few results in this direction concerns the *figure eight graph*, i.e., the graph homeomorphic to the shape of the digit 8. To state the results, let us introduce a couple of definitions and notations.

In addition to the orderings $\leq_s$, $s \geq 1$ defined in the section on n-ods (see the Definition 3.6), we need the ordering $\leq_e$ on $\mathbb{N}$ defined as follows

$$\cdots \leq_e 9 \leq_e 7 \leq_e 5 \leq_e 3,$$
$$1 \leq_e \cdots \leq_e 8 \leq_e 6 \leq_e 4 \leq_e 2, \text{ and,}$$
$$2k + 2 \leq_e k, \text{ for all odd } k \in \mathbb{N}.$$

Denote by $E(n) = \{k \in \mathbb{N} : k \leq_e n\}$ for every $n \in \mathbb{N}$. Let S_2 be an initial segment for the ordering $\leq_2$, S_3 be the union of at most two initial segments for the ordering $\leq_3$, S_4 be the union of at most three initial segments 7for the ordering $\leq_4$ and, finally, S_e be the union of at most two initial segments for the ordering $\leq_e$. We define a set A as follows. If $S_e = \{1\}$ or $S_3 \cup S_4 = \{1\}$, then $A = \emptyset$. If S_e is different from $\{1\}$ and $E(n) \cup E(n + 1)$ with $n > 2$, and $S_3 \cup S_4 \neq \{1\}$, then $A = \{2\}$. If $S_e = E(n) \cup E(n + 1)$ with $n > 2$, and $S_3 \cup S_4 \neq \{1\}$, then A is either $\emptyset$, or $\{2\}$.

Theorem 3.10 ([22]) *Let G be the figure eight graph.*

1. *If $f : G \to G$ is a continuous map fixing the branching point, then the set of periods of f is of the form $S_2 \cup S_3 \cup S_4 \cup S_e \cup A$.*
2. *Conversely, for every S of the form $S_2 \cup S_3 \cup S_4 \cup S_e \cup A$, there exists a continuous map $f : G \to G$ such that the set of periods of f is S.*

References

1. Alsedà, L., Juher, D., Mumbrú, P.: Periodic behavior on trees. Ergod. Th. & Dyn. Syst. **25**, 1373–1400 (2005)
2. Alsedà, L., Llibre, J., Misiurewicz, M.: Combinatonal Dynamics and Entropy in Dimension One. In: Advanced Series in Nonlinear Dynamics, vol. 5, 2nd edn. World Scientific, River Edge, NJ (2000)

3. Alsedà, L., Llibre, J., Misiurewicz, M.: Periodic orbits of maps of Y. Trans. Amer. Math. Soc. **313**, 475–538 (1989)
4. Baldwin, S.: An extension of Sarkovskii's Theorem to the n-od. Ergod. Th. & Dyn. Syst. **11**, 249–271 (1991)
5. Baldwin, S.: Sets of periodic points of functions on trees. In: Continuum Theory and Dynamical Systems. Contemporary Mathematics, 117, pp. 9–12. Amer. Math. Soc, Providence, RI (1991)
6. Baldwin, S., Llibre, J.: Periods of maps on trees with all branching points fixed. Ergod. Th. & Dyn. Syst. **15**, 239–246 (1995)
7. Bernhardt, C.: A Sharkovsky theorem for vertex maps on trees. J. Diff. Eq. Appl. **17**, 103–113 (2011)
8. Block, L.: Periods of periodic points of maps of the circle which have a fixed point. Proc. Amer. Math. Soc. **82**, 481–486 (1981)
9. Block, L., Guckenheimer, J., Misiurewicz, M., Young, L.-S.: Periodic points and topological entropy of one dimensional maps. In: Global Theory of Dynamic Systems. Lecture Notes in Mathematics, vol. 819, pp. 18–34. Springer, Heidelberg (1980)
10. Blokh, A. M.: On transitive mappings of one-dimensional branched manifolds (in Russian). In: Differential-Difference Equations and Problems of Mathematical Physics, pp. 3–9. Akad. Nauk Ukrain. SSR, Inst. Mat., Kiev (1984)
11. Blokh, A. M.: Dynamical systems on one-dimensional branched manifolds. I (in Russian). Teor. Funktsii Funktsional. Anal. i Prilozhen. **46**, 8–18 (1986); translation in J. Soviet Math. **48**, 500–508 (1990)
12. Blokh, A. M.: Dynamical systems on one-dimensional branched manifolds. II (in Russian). Teor. Funktsii Funktsional. Anal. i Prilozhen. **47**, 67–77 (1987); translation in J. Soviet Math. **48**, 668–674 (1990)
13. Blokh, A. M.: Dynamical systems on one-dimensional branched manifolds. III (in Russian). Teor. Funktsii Funktsional. Anal. i Prilozhen. **48**, 32–46 (1987); translation in J. Soviet Math. **49** 875–883 (1990)
14. Blokh, A. M.: On Some Properties of Graph Maps: Spectral Decomposition, Misiurewicz Conjecture and Abstract Sets of Periods. Max-Planck-Institut fuer Mathematik, Preprint #35 (1991)
15. Blokh, A.M.: The spectral decomposition for one-dimensional maps. Dyn. Report. **4**, 1–59 (1995)
16. Blokh, A.M.: On Graph-Realizable Sets of Periods. J. Diff. Eq. Appl. **9**, 343–357 (2003)
17. Bowen, R.: Periodic points and measures for Axiom A diffeomorphisms. Trans. Amer. Math. Soc. **154**, 377–397 (1971)
18. Chenciner, A., Gambaudo,J.-M., Tresser, C.: Une remarque sur la structure des endomorphismes de degré 1 du cercle. C. R. Acad. Sci. Paris, Sér. I Math. **299**, 145–148 (1984)
19. Denker, M., Grillenberger, C., Sigmund, K.: Ergodic Theory on Compact Spaces. Lecture Notes in Mathematics, vol. 527. Springer, Berlin (1976)
20. Efremova, L. S.: Periodic orbits and a degree of a continuous map of a circle (in Russian). Diff. and Integr. Equations (Gor'kii) **2**, 109–115 (1985)
21. Llibre, J., Paraños, J., Rodriquez, J.A.: Sets of periods for maps on connected graphs with zero Euler characteristic having all branching points fixed. J. Math. Anal. Appl. **239**, 85–100 (1999)
22. Llibre, J., Paraños, J., Rodriquez, J. A.: Periods for continuous self-maps of the figure-eight space. Internat. J. Bifur. Chaos Appl. Sci. Engrg. **13**, 1743–1754 (2003)
23. Misiurewicz, M.: Periodic points of maps of degree one of a circle. Ergod. Th. & Dynam. Syst. **2**, 221–227 (1982)
24. Misiurewicz, M.: Twist sets for maps of the circle. Ergod. Th. & Dynam. Sys. **4**, 391–404 (1984)
25. Poincaré, H.: Sur les courbes définies par les équations différentialles. Oeuvres completes, vol. 1, pp. 137–158. Gauthier-Villars, Paris (1952)

Chapter 4
Multidimensional Dynamical Systems

It is well known that appearance of the Sh-ordering is strongly related to the one-dimensionality of the phase space. It is natural to look for special classes of multidimensional maps (and, possibly, objects more complex than cycles) for which the Sh-ordering, or a modification thereof, take place.

To describe multidimensional maps, we use the notation

$$(x_1, \ldots, x_n) \mapsto F(x_1, \ldots, x_n) = (f_1(x_1, \ldots, x_n), \ldots, f_n(x_1, \ldots, x_n))$$

or its abbreviated version $F = (f_1, \ldots, f_n)$.

Historically, the first—and, perhaps, the most important—such class is made up of the maps such that in the above notation we have

$$f_i(x_1, \ldots, x_n) = f_i(x_1, \ldots, x_i), \quad i = 1, \ldots, n, \quad n > 1;$$

these maps are now called *triangular*. For them, the i-th coordinate depends only on the first i coordinates, i.e., a triangular map is a skew product of one-dimensional maps. In particular, iterations of triangular maps are triangular maps too. In 1979, P.Kloeden [11] proved that the Sh-ordering remains valid for triangular maps.

One more class of multidimensional maps for which some modification of the Sh-ordering holds is given by the map $F = (f_1, \ldots, f_n)$ with *cyclically permuting one-dimensional maps* of the form

$$f_i(x_1, \ldots, x_n) = f_i(x_{i+1}), \quad i = 1, \ldots, n, \quad i = i \bmod n.$$

For such a multidimensional map F, its n-th iteration $F^n = G = (g_1, g_2, \ldots)$ is transformed into the direct product of n one-dimensional maps:

$$g_i(x_1, \ldots, x_n) = f_i \circ \cdots \circ f_n \circ f_1 \circ \cdots \circ f_{i-1}(x_i).$$

© The Author(s), under exclusive license to Springer Nature Switzerland AG 2022
A. M. Blokh and O. M. Sharkovsky, *Sharkovsky Ordering*,
SpringerBriefs in Mathematics, https://doi.org/10.1007/978-3-030-99125-8_4

From this fact, one can hope to extract information about the coexistence of cycles for this type of multidimensional maps. This issue has been considered in several papers by Balibrea and Linero [6, 7].

The Sh-ordering possesses certain stability; this was first noticed by L. Block in [9]. Namely, if a one-dimensional map $f \in C^0(I, I)$ has a cycle of period k, then there exists $\varepsilon = \varepsilon(f)$ such that any map $\tilde{f} \in U_\varepsilon(f)$, with $U_\varepsilon(f)$ being the ε-neighborhood of f in $C^0(I, I)$, has cycles of every period $\tilde{k} \prec k$. This allows us to raise the question about the kindred stability of n-dimensional maps that are n-dimensional perturbations of one-dimensional maps (Zgliczynski [26]–[13, 29]).

Finally, we shall consider a special type of infinite-dimensional dynamical systems, generated by difference or difference-differential equations, and even by boundary value problems for partial differential equations.

4.1 Triangular Maps

Recall that by I we denote the unit interval $[0, 1]$, and by $I^s = [0, 1]^s$ we denote the s-dimensional unit cube. Let us look more closely at a continuous triangular map of I^n to itself denoted by

$$F = (f_1, \ldots, f_n) \quad \text{with } f_i(x_1, \ldots, x_n) = f_i(x_1, \ldots, x_i), \ i = 1, \ldots, n. \quad (4.1)$$

Theorem 4.1 ([11]) *For any $n \geq 1$, if the continuous map F (see (4.1)) has a cycle of period k, then this map also has a cycle of every period $\tilde{k} \prec k$.*

In the proof, we follow [11]. For each s, $1 \leq s \leq n$, consider the maps

$$F_s = (f_1, \ldots, f_s) : I^s \to I^s; \quad (4.2)$$

clearly, $F_1 \equiv f_1$ and $F_n \equiv F$.

Lemma 4.1 *If the map F_s, $1 \leq s \leq n - 1$, has a cycle of period $q > 1$, then the map F_{s+1} also has a cycle of period q.*

Proof Let $\bar{\beta} = (\beta_1, \ldots, \beta_s) \in I^s$ be a periodic point of the map F_s of period q. Then, for the map F_s, the points $\bar{\beta}, F_s(\bar{\beta}), \ldots, F_s^{q-1}(\bar{\beta})$ form a cycle of period q. It follows that

$$F_{s+1}^q = (h_1, \ldots, h_{s+1}); \quad (4.3)$$

the fact that $\bar{\beta}$ is a periodic point of period q for the map F_s implies that $h_i(\beta_1, \ldots, \beta_i) = \beta_i$, $i = 1, \ldots, s$ while for x_{s+1} we get that

$$h_{s+1}(\bar{\beta}, x_{s+1}) = (\bar{\beta}, H_{s+1}(x_{s+1})) \tag{4.4}$$

for the appropriately defined one-dimensional map $H_{s+1} : I \to I$. Clearly, H_{s+1} has at least one fixed point, say, γ. Then the point $(\beta_1, \ldots, \beta_s, \gamma) \in I^{s+1}$ is the desired periodic point of the map F_{s+1} of period q. $\square$

Proof of Theorem 4.1. The theorem is proved by induction. It is true for $m = 1$. Let $s \geq 1$ and suppose that the Sh-ordering takes place for the map F_s. It will be seen that this ordering also takes place for F_{s+1}.

Let η be a cycle of period $p = (2r + 1)2^l$ of the map F_{s+1}. Then F_s has a cycle $\bar{\eta}$ of period q, where q divides p. Thus, $q = (2\bar{r} + 1)2^{\bar{l}}$ with $r \geq \bar{r} \geq 0$, $l \geq \bar{l} \geq 0$, and $2\bar{r} + 1$ divides $2r + 1$. We have to consider two cases: $\bar{r} > 0$ and $\bar{r} = 0$.

1. $\bar{r} > 0$. Because $r \geq \bar{r} > 0$, $l \geq \bar{l} \geq 0$, then $(2\bar{r} + 1)2^{\bar{l}} = q \preceq p = (2r + 1)2^l$ for the map F_s. Hence, by the Sh-ordering, the map F_s has cycles of all periods

$$(2\bar{r} + 1)2^{\bar{l}} \prec \cdots \prec (2r + 1)2^l \prec (2r + 3)2^l \prec \cdots \prec 2 \prec 1.$$

   According to Lemma 4.1, in this case the map F_{s+1} also has cycles of all these periods as desired.

2. $\bar{r} = 0$. In that case $q = 2^{\bar{l}}$ with $0 \leq \bar{l} \leq l$. The map F_s has a cycle of period $2^{\bar{l}}$, and as the Sh-ordering holds for F_s by induction, the map F_s has cycles of periods

$$2^{\bar{l}-1} \prec \cdots \prec 2 \prec 1.$$

Hence, by Lemma 4.1, the map F_{s+1} also has cycles with all of these periods.

Let $\bar{\beta} = (\beta_1, \ldots, \beta_s)$ be a periodic point of the map F_s with period $q = 2^{\bar{l}}$. Consider the interval $I_{\bar{\beta}} = \{\bar{\beta}\} \times I = \{(\beta_1, \ldots, \beta_s, x), \ x \in I\}$. Since $F_{s+1}^q(\bar{\beta}) = \bar{\beta}$, the map F_{s+1}^q sends the interval $I_{\bar{\beta}}$ to itself. Moreover, the point η of F_s-period $p = (2r + 1)2^l$ has period $(2r + 1)2^{l-\bar{l}}$ under $F_{s+1}^q : I_{\bar{\beta}} \to I_{\bar{\beta}}$. Then $F_{s+1}^q|_{I_{\bar{\beta}}}$ has points of all periods t with $(2r + 1)2^{l-\bar{l}} \preceq t$. Any F_{s+1}^q-periodic point $x \in I_{\bar{\beta}}$ of F_{s+1}^q-period t is itself F_s-periodic, and its period under F_s is qt. It follows that the map F_{s+1} has points of all periods qt where t runs over the set of all numbers with $(2r + 1)2^{l-\bar{l}} \preceq t$. It is easy to see from the definition of the Sh-ordering that this covers all numbers s with $(2r + 1)2^l = p \preceq s \preceq 2^{\bar{l}}$. On the other hand, by the above the map F_{s+1} has cycles of periods

$$2^{\bar{l}-1} \prec \cdots \prec 2 \prec 1.$$

Thus, F_{s+1} has cycles of all periods r with $p \preceq r$ as desired. This completes the proof of the theorem. $\square$

Naturally, the following question comes to mind: how far can the similarity in properties of triangular maps and one-dimensional maps be extended? For example, it was discovered in S. F. Kolyada's article [12] that the topological entropy for the triangular maps, unlike that for the interval maps, may well be positive even if a triangular map has only cycles of periods 2^i, $i \geq 0$. After the appearance of [11], studying the similarity between interval maps and triangular map has been very productive, and the findings are summarized in the recent paper by M.Stefankova [24].

4.2 Cyclically Permuting Maps

Multidimensional maps $F = (f_1, \ldots, f_n)$ given by cyclically permuting one-dimensional maps of the form

$$f_i(x_1, \ldots, x_n) = f_i(x_{i+1}), \quad i = i \bmod n, \quad i = 1, \ldots, n, \qquad (4.5)$$

map the n-dimensional cube I^n to itself. They give yet another example of multidimensional maps in which the Sh-ordering plays an essential role.

For a map F, its n-th iteration $F^n = G = (g_1, g_2 \ldots)$ is transformed into the direct product of n one-dimensional maps:

$$g_i(x_1, \ldots, x_n) = f_i \circ \cdots \circ f_n \circ f_1 \circ \cdots \circ f_{i-1}(x_i).$$

For such a map F, the problem of coexistence of cycles has been investigated in detail by Balibrea and Linero. They studied not only the case of the n-dimensional cube [6] but also the n-dimensional torus T^n [7]. Here we provide selected main results of Balibrea and Linero; we follow the paper [8], in which the authors apply their findings to difference equations of the form $x_n = f(x_{n-k})$.

Balibrea and Linero define the so-called σ-*permutation* maps, with σ being a cyclic permutation of $\{1, 2, \ldots, n\}$; in fact, σ specifies how the variables are permutated. We will also use this term and will follow the authors' notations.

So, let $F : I^n \to I^n$ be the σ-permutation map defined as follows:

$$F(x_1, x_2, \ldots, x_n) = (f_{\sigma(1)}(x_{\sigma(1)}), \ f_{\sigma(2)}(x_{\sigma(2)}), \ \ldots, \ f_{\sigma(n)}(x_{\sigma(n)})).$$

Evidently, the set of periods of cycles for the σ-permutation map is closely related to that for the one-dimensional maps $f_i^{(n)} : I \to I$, $i = 1, 2, \ldots, n$, given by

$$f_i^{(n)} = f_{\sigma(i)} \circ f_{\sigma^2(i)} \circ \cdots \circ f_{\sigma^n(i)}.$$

We denote by $\text{Per}(f)$ the set of periods of cycles of f and by $S(m)$ the initial segment of the Sh-ordering which ends at $m \in \mathbb{N}^* = \mathbb{N} \cup 2^\infty$, that is,

$$S(m) = \{n \in \mathbb{N}^* : n \prec m\} \quad \text{if } m \in \mathbb{N}$$

and

$$S(2^\infty) = \{1, 2, 2^2, \ldots, 2^n, \ldots\}.$$

For any continuous interval map f, Sh-Theorem establishes that there exists $m \in \mathbb{N}^*$ such that $\text{Per}(f) = S(m)$.

Let $\gcd(p, q)$ denote the greatest common divisor of the positive integers p, q, and $p|q$ mean that p is a divisor of q. For $n \in \mathbb{N}$ and $m \in \mathbb{N}^*$ we introduce

$$S_n(m) = \left\{ t \in \mathbb{N} : t \nmid n \text{ and } \frac{t}{\gcd(t, n)} \in S(m) \right\} \cup \{1\}.$$

Obviously, $t \geq m$ implies $S_n(m) \subseteq S_n(t)$. Moreover, it is easy to see that

$$S_n(m) = \left\{ pt : p|n, \ t \in S(m) \setminus \{1\}, \ \gcd\left(\frac{n}{p}, t\right) = 1 \right\} \cup \{1\}.$$

Theorem 4.2 (Periodic structure of σ-permutation maps on I^n [6, 8]) *For any σ-permutation map $F : I^n \to I^n$, there exists $m \in \mathbb{N}^*$ such that*

$$\text{Per}(F) = S_n(m) \quad or \quad Per(F) = S_n(m) \cup \{p : p|n\}, \tag{4.6}$$

and vise versa, for any $m \in \mathbb{N}^*$, there exists a σ-permutation map $F : I^n \to I^n$ such that (4.6) is valid. Moreover, $\text{Per}(F) \neq S_n(m)$ if and only if $f_i^{(n)}$ has at least two fixed points.

The key to the proof of the theorem is to use the periodic structure of the interval maps $f_i^{(n)}$ (see [6]). By periodic structure of a map, we mean a type of forcing relation among all the periods that the map has. Theorem 4.2 can be extended to the case of $\mathbb{R}^n$ leading to similar results. The arguments can be successfully adapted to $\mathbb{R}^n$ except for the situation $\text{Per}(F) = \emptyset$, which gives another (trivial) possibility for both the direct and converse statements.

In order to better visualize the above periodic structure, we consider the cases $n = 2, 3$. We use the notation $a \Rightarrow b$ to indicate that the presence of period a forces the presence of period b in a map, and $a \Leftrightarrow b$ indicates that periods a and b are mutually forced.

Theorem 4.3 (Periodic structure on I^2 [6, 8]) *The periodic structure of a σ-permutation map $F : I^2 \to I^2$ is described in the frame of forcing on $\mathbb{N} \setminus \{2\}$ by the ordering*

$$(3 \Leftrightarrow 2 \cdot 3) \Rightarrow (5 \Leftrightarrow 2 \cdot 5) \Rightarrow \cdots \Rightarrow (2n + 1 \Leftrightarrow 2(2n + 1)) \Rightarrow \cdots$$

$$\Rightarrow 2^2 \cdot 3 \Rightarrow 2^2 \cdot 5 \Rightarrow \cdots \Rightarrow 2^2(2n + 1) \Rightarrow \cdots$$

$$\Rightarrow 2^3 \cdot 3 \Rightarrow 2^3 \cdot 5 \Rightarrow \cdots \Rightarrow 2^3(2n + 1) \Rightarrow \cdots$$

$$\Rightarrow 2^k \cdot 3 \Rightarrow 2^k \cdot 5 \Rightarrow \cdots \Rightarrow 2^k(2n + 1) \Rightarrow \cdots$$

$$\Rightarrow 2^{m+1} \Rightarrow 2^m \Rightarrow \cdots \Rightarrow 2^3 \Rightarrow 2^2 \Rightarrow 1 .$$

Moreover, $2 \in Per(F)$ *if and only if* $f_i^{(2)}$ *has at least two fixed points.*

Theorem 4.4 (Periodic structure on I^3 [6, 8]) *The periodic structure of a* σ-*permutation map* $F : I^3 \to I^3$ *is described in the frame of forcing on* $N \setminus \{3\}$ *by the ordering*

$$3 \cdot 3 \Rightarrow (3 \cdot 5 \Leftrightarrow 5) \Rightarrow (3 \cdot 7 \Leftrightarrow 7) \Rightarrow 3 \cdot 9 \Rightarrow (3 \cdot 11 \Leftrightarrow 11) \Rightarrow \cdots$$
$$\Rightarrow 3 \cdot 2 \cdot 3 \Rightarrow (3 \cdot 2 \cdot 5 \Leftrightarrow 2 \cdot 5) \Rightarrow (3 \cdot 2 \cdot 7 \Leftrightarrow 2 \cdot 7) \Rightarrow 3 \cdot 2 \cdot 9 \Rightarrow \cdots$$
$$\Rightarrow 3 \cdot 2^k \cdot 3 \Rightarrow (3 \cdot 2^k \cdot 5 \Leftrightarrow 2^k \cdot 5) \Rightarrow (3 \cdot 2^k \cdot 7 \Leftrightarrow 2^k \cdot 7) \Rightarrow 3 \cdot 2^k \cdot 9 \Rightarrow \cdots$$
$$\Rightarrow (3 \cdot 2^m \Leftrightarrow 2^m) \Rightarrow \cdots \Rightarrow (3 \cdot 2^2 \Leftrightarrow 2^2) \Rightarrow (3 \cdot 2 \Leftrightarrow 2) \Rightarrow 1 .$$

Moreover, $3 \in Per(F)$ *if and only if* $f_i^{(3)}$ *has at least two fixed points.*

Consider now the periodic structure of σ-permutation maps on the n-dimensional torus $\mathbb{T}^n = \underbrace{\mathbb{S}^1 \times \cdots \times \mathbb{S}^1}_{n}$ with $n \geq 2$ and $\mathbb{S}^1$ being the unit circle $\{z \in \mathbb{C} : |z| = 1\}$.
We need certain results on circle maps (a complete treatment of the subject can be found in [1]). Here the pivotal notion is that of the degree of a σ-permutation map F on $\mathbb{T}^n$, denoted by D and defined as

$$D = deg(F) = deg(f_i^{(n)}) = \prod_{i=1}^{n} deg(f_i) .$$

The sought-for periodic structure depends on the periodic structure of the maps $f_i^{(n)}$, $i = 1, \ldots, n$, which in turn depends on the degree of the map. Thus, the periodic structure of F is influenced by D. We separate the cases: $D \neq 1$ and $D = 1$. In the former, we will again need the notion of initial segment $S_n(m)$ of the Sh-ordering, that was introduced in the previous case.

Theorem 4.5 ([7, 8]) *Let* $F : \mathbb{T}^n \to \mathbb{T}^n$ *be a* σ-*permutation map.*

(1) If $D = 0$, *then there exists* $m \in \mathbb{N}^*$ *such that* $Per(F) = S_n(m)$ *or*
$$S_n(m) \cup \{p : p|n\}.$$
(2) If $D = -1$, *then there exists* $m \in \mathbb{N}^*$ *such that* $Per(F) =$
$$= S_n(m) \cup \{p : p|n\}.$$
(3) If $D = 2$, *then* $Per(F) = \mathbb{N}$ *or* $Per(F) = \mathbb{N} \setminus \{p : p|n, p \geq 2\}$.
(4) If $D = -2$, *then* $Per(F) = \mathbb{N}$ *or* $Per(F) = \mathbb{N} \setminus \{2p : p|n, 2p \nmid n\}$.
(5) If $|D| \geq 2$, *then* $Per(F) = \mathbb{N}$.

For the case $D = 1$ we refer the reader to [8].

4.3 Multidimensional Perturbations of One-Dimensional Maps

In studying mathematical models of various phenomena, it is important (in particular, from the physical point of view) to consider properties that are stable in a certain sense under perturbations.

For most of stability theorems, it suffices to consider C^1-maps and at least C^1-perturbations. Louis Block [9] was the first to notice that the Sh-ordering for maps $f \in C^0(I, I)$ has some stability even in the C^0-topology, i.e., with respect to all maps from $C^0(I, I)$ close enough to f.

Theorem 4.6 *If a map $f \in C^0(I, I)$ has a cycle of period m then there exists some $\varepsilon = \varepsilon(f) > 0$ such that every map $g \in C^0(I, I)$ from the ε-neighborhood of f has a cycle of any period k if $k \prec m$.*

This fact, generally speaking, is a consequence of the lower semicontinuity of the map $x \mapsto orbit(x, f)$ for any $f \in C^0(I, I)$.

There are a lot of other versions of the property that could be called C^0-stability of Sh-ordering. For example, Zgliczynski used this idea many times in computer-assisted proofs of chaos in different models (see, in particular, [30]) or when analyzing multidimensional perturbations of one-dimensional maps ([26]–[28]). Here we only consider some of these applications. In our presentation we follow [26, 28].

Let $f : \mathbb{R} \to \mathbb{R}$ be a continuous map. Let V be a real Banach space and let a continuous decomposition $V = \mathbb{R} \oplus W$ be given. According to this decomposition, we will represent elements $v \in V$ as pairs $v = (x, w)$, where $x \in \mathbb{R}$ and $w \in W$. Let $F : [0, 1] \times V \to V$ be a continuous and compact map. In the sequel we will use the notation F_λ for the partial map with fixed $\lambda \in [0, 1]$, so $F_\lambda : V \to V$ is defined as $F_\lambda(v) := F(\lambda, v)$. Suppose that $F_0(x, w) = (f(x), 0)$. We say that the maps F_λ are *multidimensional perturbations* of f.

In [26, 28] it was shown that the Sh-ordering is stable with respect to multidimensional perturbations of f.

Theorem 4.7 ([28]) *Let $f : \mathbb{R} \to \mathbb{R}$ be continuous and $F : [0, 1] \times V \to V$ be continuous with $F_0(x, w) = (f(x), 0)$. If f has a periodic point of period k, then for any $r > 0$, there exists $\lambda_0 > 0$ such that for all $0 \leq \lambda \leq \lambda_0$ and all $m \prec k$, the map F_λ has a periodic point of period m in the set $\mathbb{R} \oplus B_n(r)$, where $B_n(r) \subset \mathbb{R}^n$ is an open ball of radius r centered at the origin in $\mathbb{R}^n$.*

The proof of the theorem uses the notion of covering relation in multidimensional situation and the continuation of one-dimensional orbits with non-zero fixed point index for multidimensional perturbations of one-dimensional maps (see, for instance, [10]).

Recently, due to biological challenges and, in particular, due to the desire to understand how the brain works or can work, the research into chains of coupled oscillators (for example, coupled neurons) becomes very popular. Consider, for example, a continuous map $F_\lambda : \mathbb{R}^n \to \mathbb{R}^n$ given by

$$F_\lambda(x_1, \ldots, x_n) = (f_1(x_1), f_2(x_2), \ldots, f_n(x_n)) + \lambda\, G(x_1, \ldots, x_n). \qquad (4.7)$$

Lemma 4.2 *Let* $F_\lambda : \mathbb{R}^n \to \mathbb{R}^n$ *be of the form (4.7). If each of one-dimensional maps* $f_i : \mathbb{R} \to \mathbb{R}$, $i = 1, \ldots, n$, *has, respectively, a periodic point of period* k_i, *then there exists* $\lambda_0 > 0$ *such that for all* $0 \le \lambda \le \lambda_0$ *and all numbers* m *with* $m \prec k_i$, $i = 1, \ldots, n$, *the map* F_λ *has a periodic point of period* m.

This lemma follows from Theorem 4.7 as applied to f_i, $i = 1, \ldots, n$. Indeed, by the theorem, given m satisfying conditions of the theorem, for each $i \in \{1, \ldots, n\}$ there exists λ_i such that for all $0 \le \lambda \le \lambda_i$ the map f_i has a periodic point of period m. Now, it remains to put $\lambda_0 = \min\{\lambda_1, \ldots, \lambda_n\}$.

Another related paper is [29] in which Zgliczynski studies periodic points for systems of weakly coupled one-dimensional maps. The paper considers the map $F : \mathbb{R}^n \to \mathbb{R}^n$ given by

$$F(x_1, \ldots, x_n) := (f(x_1), f(x_2), \ldots, f(x_n)) + G(x_1, \ldots, x_n)$$

with $f(x) = 1 - \mu x^2$, μ is close to 2, and G is a small perturbation.

The author shows that if G is sufficiently small in C^0-norm, then the number of periodic points of the map F is infinite and there exists a phase space region such that every ball in this region that is not too small and contains a periodic point of F. He also gives an estimate for G.

In the paper by Li and Zgliczynski [13] devoted to the stability of forcing relations for multidimensional perturbations of interval maps, the above theorem, regarding the forcing relations for cycle's periods, is extended to those for cycle's patterns. To state its results we need several definitions; even though some of them have already been given earlier, for convenience, we repeat them here [1].

By a *cycle*, we will mean a pair (P, φ) with $P \subset \mathbb{R}$ being a finite non-empty set and φ being a cyclic permutation of P. The number of elements in P will be called the *period* of (P, φ) and will be denoted by $|P|$. If f is a continuous self-mapping of $\mathbb{R}$ and (P, φ) is a cycle, we say that f has the cycle P, if $\varphi = f|_P$, the restriction of f to P.

Let $\langle S \rangle$ be for the smallest closed interval containing the set $S \subset \mathbb{R}$. The cycles (P, φ) and (Q, ψ) are declared to be equivalent if (and only if) there exists a homeomorphism $h : \langle P \rangle \mapsto \langle Q \rangle$ such that $h(P) = Q$ and $\psi \circ h|_P = h \circ \varphi|_Q$. Equivalence classes in the set of all cycles in R will be called *patterns*.

The *forcing relation* between patterns is defined as follows: with $\mathscr{A}$ and $\mathscr{B}$ being two patterns, we say that $\mathscr{A}$ *forces* $\mathscr{B}$ and write $(\mathscr{A} \overset{pat}{\Rightarrow} \mathscr{B})$ if and only if every continuous map on R which has a cycle with pattern $\mathscr{A}$ has a cycle with pattern $\mathscr{B}$. For a pattern $\mathscr{A}$, we put $Per(\mathscr{A}) = \{|\mathscr{B}| : \mathscr{B} \ne \mathscr{A}$ is a pattern such that $\mathscr{A} \overset{pat}{\Rightarrow} \mathscr{B}\}$.

With these definitions, we can formulate the main result: all periods of periodic points forced by a pattern for interval maps are preserved for high-dimensional maps if the multidimensional perturbation is small enough.

Theorem 4.8 ([13]) *Let $F : \mathbb{R} \times V \to V$, $V = \mathbb{R} \oplus \mathbb{R}^n$, be a continuous map such that its associated partial map $F_\lambda : V \to V$, $\lambda \in \mathbb{R}$, has the property: $F_0(x, y) = (f(x), g(x))$ for all $(x, y) \in V$, where f is continuous on $\mathbb{R}$ and g is either a continuous map from $\mathbb{R}$ to $\mathbb{R}^n$, or g is continuous on $\mathbb{R} \oplus S$ and $g(\mathbb{R} \oplus S) \subset int(S)$ with $S \subset \mathbb{R}^n$ being some compact set homeomorphic to the closed unit ball in $\mathbb{R}^n$. If f has a pattern $\mathscr{A}$, then there exists $\lambda_0 > 0$ such that for $\lambda < \lambda_0$ the map F_λ has periodic points of all periods from $Per(A)$.*

Thus, all cycle's periods forced by a pattern for interval maps are preserved for the above class of multidimensional maps when perturbations are small.

4.4 Infinitely-Dimensional Dynamical Systems, Generated by One-Dimensional Maps

It is known that some problems in mathematical physics can be reduced to difference equations of the form

$$x(t + 1) = f(x(t)), \quad t \in \mathbb{R}, \tag{4.8}$$

i.e., here the time is continuous. For example, the simplest boundary value problem $w_t - w_y = 0$, $y \in [0, 1]$, $t \in \mathbb{R}^+$, $w|_{y=1} = f(w|_{y=0})$ is reduced directly to the equation $u(t + 1) = f(u(t))$ inasmuch as the general solution of the partial differential equation has the form $w(t, y) = u(t + y)$ with u being an arbitrary function and its substitution into the boundary condition results in the difference equation.

A bit more complicated boundary value problem $w_t - w_y = 0$, $y \in [0, 1]$, $t \in \mathbb{R}^+$, $\frac{\partial w}{\partial y}|_{y=1} = g(w|_{y=0}) \frac{\partial w}{\partial y}|_{y=0}$, is reduced to the differential-difference equation $u'(\tau + 1) = f(u(\tau)) u'(\tau)$, which, after integration, gives the family of difference equations $u(\tau + 1) = f(u(\tau)) + \lambda$, where f is an antiderivative of g and $\lambda = w(0, 1) - f(w(0, 0))$. Some reducible boundary value problems for the wave equation can be found in [16, 17]. These examples suggest that the study of sufficiently smooth solutions to (4.8) is quite important.

What can we expect of the solutions to the continuous-time difference Eq. (4.8)? Do these have any radically new properties compared to the properties of the solutions to the similar discrete-time equations $x_{n+1} = f(x_n)$? Every solution of (4.8) is determined by its values on the interval $0 \le t < 1$ (and not its value at $t = 0$, as in the case of discrete time) and hence Eq. (4.8) generates the infinite-dimensional dynamical system $\varphi \mapsto f(\varphi)$ on the space of initial functions $\varphi : [0, 1) \to \mathbb{R}$. And now that the points of the phase space are functions, we must speak about the spatial-temporal behavior of trajectories, about the evolution of functions φ with time (when they are "moving along trajectories"). In this case, as would be expected, all complexities in the temporal behavior of trajectories of one-dimensional dynamical systems $x \mapsto f(x)$ are transformed into a very intricate evolution of functions φ.

To every solution of Eq. (4.8), there corresponds a continual family of trajectories of the one-dimensional map $x \mapsto f(x)$. Namely, at every point $t_* \in [0, 1)$, the oscillator $\varphi(t_*) \mapsto f(\varphi(t_*))$ acts; its oscillations are independent of the oscillators located at other points of the interval $[0, 1)$, and hence Eq. (4.8) can be interpreted as the problem about the dynamics of a continuum of uncoupled oscillators. Mere independence of the oscillations may lead to solutions whose large-time behavior is virtually indistinguishable from that of random processes. This phenomenon, which was called *self-stochasticity* [18], implies, when applied to boundary value problems, the self-generation and development of spatial-temporal chaos. Observe that the study of spatial-temporal chaos needs probability methods (see, e.g., [14]). Another important consequence of the above-mentioned relation between the behavior of the solutions to Eq. (4.8) and the dynamics of the map $x \mapsto f(x)$ is the *cascading of coherent structures* in solutions of appropriate boundary problems (see, e.g., [15]). Such self-structuring phenomena are consequences of the complicated geometry of attraction basins of cycles and cycles of intervals of one-dimensional maps [19].

Here, for the sake of simplicity, we restrict ourselves to the case of quadratic f, also bearing in mind that there is much more information about quadratic maps than general maps. What kind of solutions of Eq. (4.8) should we speak about then ? Here, we do not mean the piecewise constant solutions, which merely reflect (repeat) the dynamics of f. It is expedient to regard as solutions the continuous functions and, besides these, upper semicontinuous functions, if we want to study the asymptotic behavior of continuous solutions.

Every solution of Eq. (4.8) can be represented in the form

$$x(t) = f^n(\varphi(t - n)), \quad \text{if} \quad n \le t < n + 1, \quad n = 1, 2, \ldots \quad (4.9)$$
$$\text{where} \quad \varphi(t) = x(t)|_{[0,1)}.$$

Hence a solution $x(t)$ is continuous if $f(x)$ and its associated initial function $\varphi(t)$, $t \in [0, 1)$, are continuous and satisfy the consistency condition $\varphi(1 - 0) = f(\varphi(0))$. Thus, in order to understand how the solutions could behave, it is necessary to know how the iterations of $f(x)$ behave. Below we list some typical kinds of limiting behavior that $f^n(x)$ can have when n increases; as an example we consider nonlinear functions $f_\lambda(x) = \lambda - x^2$. To do this we use, along with the Hausdorff metric, the concepts of basin of attraction of an attracting cycle and of the domain of influence of a repelling cycle. Recall that the latter is defined by $Q(\Gamma) = \cap_{\varepsilon>0} \cap_{j>0} \cup_{n>j} f_\lambda^n(V_\varepsilon(\Gamma))$ with Γ being a repelling cycle and $V_\varepsilon(\cdot)$ being the ε-neighborhood of a set.

When $\lambda \in [-1/4, 2)$, the map $f_\lambda(x)$ has the invariant interval $I_\lambda = [-a, a]$, $a = \frac{1+\sqrt{1+4\lambda}}{2}$, and $f_\lambda(x)$, being a quadratic function, can have no more than one attracting cycle or cycle of intervals.

For smooth 1D maps, two types of bifurcations are typical: period-doubling bifurcation (soft bifurcation), in which an attracting cycle becomes repelling and gives birth to an attracting cycle with double the period of the original cycle, and tangent bifurcation (hard bifurcation), in which there arise two cycles (of which one is attracting and the other is repelling) and a cycle of intervals (containing these cycles), all three having the same period, and wherein the repelling cycle and its preimages constitute the boundary of the cycle of intervals. In families of quadratic maps, only such bifurcations of cycles are possible (at that (notice that a tangent bifurcation may happen only after an infinite number of period-doubling bifurcations).

It is essential that from the moment of tangent bifurcation and with a further increase of λ, the qualitative changes in the behavior on $I_\lambda = [-a, a]$ come down to the changes only on the cycle of intervals $J_{(m)}$, $m \geq 1$, that results from the tangent bifurcation. The evolution of $f_\lambda(x)$ on $J_{(m)}$ follows the same pattern that is observed on the interval $I_\lambda = [-a, a]$, with the only difference being that the changes occur on the m intervals of $J_{(m)}$ simultaneously and, in terms of λ, as long as these intervals are invariant with respect to f_λ^m.

Thus, we can identify the closed parameter interval $\Lambda_{(m)} \subset [-1/4, 2]$ whose endpoints are defined by two values of λ, of which one is the value corresponding the birth of two period-m cycles and a period-m cycle of intervals (due to the associate tangent bifurcation) and the other is the value at which this cycle of intervals disappears, i.e., its intervals become non-invariant with respect to f_λ^m (just like the interval I ceases to be invariant with respect to f when $\lambda > 2$). In the case of period-doubling bifurcation, the situation is similar, with the difference being that a cycle of intervals of period $2m$ arises (one can say that a periodic orbit "blows up" and creates the cycle of intervals in question).

The following can be concluded from the above. If the map f_λ has an attracting period-m cycle $\mathbb{A}_\lambda = \{\alpha_1, \ldots, \alpha_m\}$ then there is a finite sequence of integers-preperiods $m_1, m_2, \ldots, m_k$ (that shows the prehistory of the emergence of $\mathbb{A}_\lambda$) with the listed below properties: 1) $m = \Pi_{i=1}^{k} m_i$ and $m_i > 2$ for every $i \leq k$ or there exists j, $1 \leq j \leq k$, such that $m_i = 2$ for all $i \geq j$, 2) there exists a sequence of nested cycles of intervals $J_{s_1} \supset J_{s_2} \supset \ldots \supset J_{s_k}$ with periods $s_i = m_1 \cdot \ldots \cdot m_i$, $i = 1, \ldots, k$, containing $\mathbb{A}_\lambda$. Wherein each of the s_i intervals constituting J_{s_i} contains m_{i+1} intervals from $J_{s_{i+1}}$ and the boundary of J_{s_i} is formed by the (unique) repelling cycle of period s_i, say, $\Gamma_i = \{\gamma_i^1, \gamma_i^2, \ldots, \gamma_i^{s_i}\}$, and its preimages (or is formed by the repelling cycle of period $2s_i$, if $m_i = 2$); and wherein every interval from J_{s_i} contains only one point of Γ_i. Let Q_i^j be the interval from J_{s_i} which contains the point γ_i^j. The union of the intervals Q_i^j, $j = 1, \ldots, s_i$, gives the domain of influence $Q(\Gamma_i)$ of the cycle Γ_i and, as already mentioned, $Q_i^j \supset Q_{i+1}^{r \bmod m_i}$, $i = 1, \ldots, k-1$, $j = 1, \ldots, s_i$, $r = 1, \ldots, s_{i+1}$.

Besides, there exists a sequence of nested intervals $[-1/4, 2] \supset \Lambda_{s_1} \supset \Lambda_{s_2} \supset \ldots \supset \Lambda_{s_k}$ such that for every $\lambda \in \Lambda_{s_k}$ the map f_λ has an attracting period-m cycle $\mathbb{A}_\lambda$ with the same, mentioned above, properties.

The basin of attraction of the attracting cycle $\mathbb{A}_\lambda$ consists of m subsets B_i, $i = 1, \ldots, m$, and each of B_i consists of a countable number of disjoint intervals such that $B_i \ni \alpha_i$ and $f_\lambda(B_i) = B_{i+1}$, $i = 1, \ldots, m$, $i = i \bmod m$. It is known, that $\cup_{i=1}^{m} B_i = I_\lambda$.

There exists an upper semicontinuous function $f_\lambda^*(x)$ whose graph is the topological limit of the graphs of the iterations $f_\lambda^{mn}(x)$, $x \in I_\lambda$, $n = 1, 2, \ldots$, such that

$$f_\lambda^*(x) = \begin{cases} \alpha_i, & \text{if } x \in B_i, \\ Q_i^j, & \text{if } x \in \overline{\cup_{n=0}^\infty f_\lambda^{-n}(\gamma_i^j)}, \quad j = 1, \ldots, s_i, \quad i = 1, \ldots, m. \end{cases}$$

Hence for the dynamical system induced by Eq. (4.8) on the continuous functional space completed (in the Hausdorff metric for graphs) with upper semicontinuous functions, the ω-limit set of the trajectory starting at $\varphi : [0, 1] \to I_\lambda$ is the period-m cycle consisting of the upper semicontinuous functions $\varphi^*(t)$, $f_\lambda(\varphi^*(t))$, $\ldots$, $f_\lambda^m((\varphi^*(t)))$, where $\varphi^*(t) = f_\lambda^*(\varphi(t))$, $t \in [0, 1]$. Consequently, the corresponding solution of (4.8) approaches (in the Hausdorff metric for graphs) the period m upper semicontinuous function $P(t) = f_\lambda^{[t]}(\varphi^*(\{t\}))$, $t \in \mathbb{R}^+$, with $[\cdot]$ and $\{\cdot\}$ standing for the integer and fractional parts of a number, respectively. This fact allows us to formulate the following theorem.

Theorem 4.9 ([20]) *If the map f_λ at $\lambda = \lambda_1$ has no cycles of period n_1 and at $\lambda = \lambda_2$ has a cycle of period n_2, and $n_1 \prec n_2$, then for any n with $n_1 \prec n \prec n_2$, there exists an open interval $\Lambda_n \subset [-1/4, 2]$ such that for $\lambda \in \Lambda_n$ almost every bounded solution of (4.8) is asymptotically periodic with period n.*

Since for the unimodal maps, it is possible also to use the rotation numbers of cycles $\rho = p/m$ (where m is the period of cycle and p is the number of its elements which are mapped to the left of themselves), we can complete Theorem 4.9.

Theorem 4.10 *The map f_λ at $\lambda = \lambda_1$ has a cycle with the rotation number ρ_1 and at $\lambda = \lambda_2$ has no cycles with the rotation number ρ_2 and $\rho_1 > \rho_2$, then for any rotation pair (p, m) with $\rho_1 > p/m > \rho_2$, there exists an open interval $\Lambda_{p,m} \subset [-2, 1/4]$ such that for any $\lambda \in \Lambda_{p,m}$ f_λ has an attracting cycle with the rotation pair (p, m) and almost each bounded solution of (4.8) is asymptotically m-periodic with the rotation pair (p, m), and the ω-limit set of the trajectory corresponding to this solution is a periodic trajectory of period m in the space C^Δ.*

It should be noted that in above theorems, we can replace $f(x) = \lambda - x^2$ by an arbitrary unimodal convex function $f(x, \lambda)$ with $\frac{df}{d\lambda} > 0$ and negative Schwarzian derivative. So, here we can speak about the stratification of the smooth 1-D maps space depending on the asymptotic behavior of solutions of corresponding difference equations (with continuous time).

4.5 Final Remarks

In the end of this chapter, we want to briefly discuss several papers further developing the ideas and methods related to the Sh-theorem.

4.5.1 Multivalued Maps

A class of maps for which an analog of the Sh-theorem holds are the so-called *multivalued* maps. This class of maps is studied in several papers by Andres, Pastor and their collaborators. To give the reader a flavor of what is proven in these papers, we describe the results of [3] relevant to the Sh-theorem. The authors define a class of so-called *M-functions*, i.e. multi-valued functions $\Phi : \mathbb{R} \to \mathbb{R}$ that have the following properties: (a) for any x, the set $\Phi(x)$ of values of Φ at x is either a point or a closed interval, and (2) the function Φ is upper-semicontinuous. The authors define periodic orbits for an M-map ϕ as follows: an n-orbit is a sequence $\{x_i\}_{i=0}^{i} nfty$ such that (i) $x_{i+1} \in \phi(x_i)$ for $i \geq 0$, (ii) $x_i = x_{i+n}$ for $i \geq 0$, and (iii) the string $x_0, x_1, \ldots, x_{n-1}$ is not repetitive. The authors prove that the Sh-theorem holds for M-maps with at most two exceptions. They then interpret this in terms of differential equations and difference enclosures. These ideas are then developed in later articles by the same authors and their collaborators [4, 5].

4.5.2 Nonlinear Difference Equations

In the end of this chapter, we would like to mention one direction related to the Sh-theorem in which dynamical systems theory has been developing for a while. We mean the *difference equations*, a discrete analog of ordinary differential equations. Even though this is a nice topic closely related to one-dimensional dynamics (see [21]), including it in the current book would increase its size beyond the presumed scope. Thus, we decided to simply refer the reader to the book [21] and to the references therein. As an example of recent developments here, we also would like to mention the paper [2] in which the Sh-theorem is extended onto so-called *periodic difference equations*.

References

1. Alsedà, L., Llibre, J., Misiurewicz, M.: Combinatonal dynamics and entropy in dimension one. In: Advanced Series in Nonlinear Dynamics, vol. 5, 2nd edn. World Scientific, River Edge, NJ (2000)
2. AlSharawi, Z., Angelos, J., Elaydi, S., Rakesh, L.: An extension of Sharkovsky's theorem to periodic difference equations. J. Math. Anal. Appl. **316**(1), 128–141 (2006)
3. Andres, J., Pastor, K.: A version of Sharkovskii's theorem for differential equations. Proc. Amer. Math. Soc. **133**(2), 449–453 (2005)
4. Andres, J., Pastor, K., Šnyrychová, P.: A multivalued version of Sharkovskii's theorem holds with at most two exceptions. J. Fixed Point Theory Appl. **2**(1), 153–170 (2007)
5. Andres, J., Fürst, T., Pastor, K.: Full analogy of Sharkovsky's theorem for lower semicontinuous maps. J. Math. Anal. Appl. **340**(2), 1132–1144 (2008)
6. Balibrea, F., Linero, A.: Periodic structure of σ-permutation maps on I^n. Aequationes Math. **62**(3), 265–279 (2001)

7. Balibrea, F., Linero, A.: Periodic structure of σ-permutation maps. II. The case T^n. Aequationes Math. **64**(1-2), 34-52 (2002)
8. Balibrea, F., Linero, A.: On the periodic structure of delayed difference equations of the form $x_n = f(x_{n-k})$ on I and S^1. J. Diff. Eq. Appl. **9**(3–4), 359–371 (2003)
9. Block, L.: Stability of periodic orbits in the theorem of Sharkovskii. Proc. Amer. Math. Soc. **81**, 333–336 (1981)
10. Gidea, M., Zgliczynski, P., : Covering relations for multidimensional dynamical systems. I, II. J. Diff. Eq. Appl. **202**, 32–58 (2004)
11. Kloeden, P.E.: On Sharkovsky's cycle coexisting ordering. Bull. Austral. Math. Soc. **20**, 171–177 (1979)
12. Kolyada, S.F.: On dynamics of triangular maps of the square. Ergod. Th. & Dyn. Syst. **12**, 749–768 (1992)
13. Li, M.-Ch., Zgliczynski, P.: On stability of forcing relations for multidimensional perturbations of interval maps. Fund. Math. **206**, 241–251 (2009)
14. Romanenko, EYu.: Randomness in deterministic continuous-time difference equations. J. Diff. Eq. Appl. **16**(2–3), 243–268 (2010)
15. Romanenko, EYu., Sharkovsky, A.N., Vereikina, M.B.: Self-structuring and self-similarity in boundary value problems. Int. J. Bifurcation & Chaos **5**(5), 1407–1418 (1995)
16. Sharkovsky, O. M.: Dynamical systems generated by boundary value problems. Ideal turbulence. Computer turbulence (in Ukrainian). In: Proceedings of Ukrainian Math. Congress, Vol. 12, pp. 125–149. Kiev, Ukraine (2003)
17. Sharkovsky, A.N.: Difference equations and boundary value problems. In: New Progress in Difference Equations. Proceedings of ICDEA 2001, pp. 3–22. Taylor & Francis, Boca Raton, FL (2004)
18. Sharkovsky, A.N., Romanenko, EYu.: Ideal turbulence: Attractors of deterministic systems may lie in the space of random fields. Int. J. Bifurcation & Chaos **2**(1), 31–36 (1992)
19. Sharkovsky, A.N.: Ideal turbulence. Nonlinear Dynamics **44**, 15–27 (2006)
20. Sharkovsky, A.N.: Universal phenomena in some infinite-dimensional dynamical systems. Int. J. Bifurcation & Chaos **5**(5), 1419–1425 (1995)
21. Sharkovsky, A. N.; aĭstrenko, Yu. L.; Romanenko, E. Yu.: Difference equations and their applications. Translated from the 1986 Russian original by D. V. Malyshev, P. V. Malyshev and Y. M. Pestryakov. Mathematics and its Applications, 250. Kluwer Academic Publishers Group, Dordrecht (1993)
22. Sharkovsky, A.N., Sivak, A.G.: Universal order and universal rate of solution bifurcations of difference-differential equations (in Russian). In: Approximate and Qualitative Methods of the Theory of Differential and Functional-Differential Equations, pp. 98–106. Inst. Mat., Ukrain. Akad. Nauk, Kiev (1983)
23. Sharkovsky, A.N., Sivak, A.G.: Universal phenomena in solution bifurcations of some boundary value problems. Nonlinear Math. Physics **1**(2), 147–157 (1994)
24. Stefankova, M.: The Sharkovsky program of classification of triangular maps - A Survey. Topol. Proc. **48**, 135–150 (2016)
25. Zgliczynski, P.: Fixed point index for iterations of maps, topological horseshoe and chaos. Topological Methods in Nonlinear Analysis **8**(1), 169–177 (1996)
26. Zgliczynski, P.: Sharkovskii's theorem for multidimensional perturbations of one-dimensional maps. Ergod. Th. & Dyn. Syst. **19**(6), 1655–1684 (1999)
27. Zgliczynski, P.: Sharkovskii's theorem for multidimensional perturbations of one-dimensional maps II. Topological Methods in Nonlinear Analysis **14**, 169–182 (1999)
28. Zgliczynski, P.: Multidimensional perturbations of one-dimensional maps and stability of Sharkovskii ordering. Int. J. Bifur. & Chaos **9**(9), 1867–1876 (1999)
29. Zgliczynski, P.: On periodic points for systems of weakly coupled 1-dim maps. Nonlinear Anal. TMA **46**(7), 1039–1062 (2001)
30. Zgliczynski, P.: Computer assisted proof of chaos in the Henon map and in the Rossler equations. Nonlinearity **10**, 243–252 (1997)

Chapter 5
Historical Remarks

In the end of 1989 at the invitation of George R. Sell (University of Minnesota), I visited the US for the first time and spent 2 months in Minneapolis. Before coming back to Kiev I gave a talk at John Milnor's seminar and then, as did a number of mathematicians who visited New York City at that time, stayed for a few days at Dennis Sullivan's apartment in New York City, and heard from my host a phrase that sounded more or less as follows: Lived as all others, but all of a sudden woke up famous.

However, in reality, the progress towards the ultimate goal—the publication of the article "Coexistence of the cycles of a continuous map of the line into itself" in the Ukrainian Mathematical Journal [UMZh, 1964, 16, No. 1, 61–71]—was more or less traditional and lasted about 2 years. According to the dates listed in the journal publications' record, it began in May 1960 with the statement $\forall(k > 2) \succ 2$ which appeared in the article "Necessary and sufficient conditions for the convergence of one-dimensional iterative processes" [UMZh, 1960, 12, No. 4], then continued with the statement $\forall k \neq 2^i \succ ... 2^m \succ 2^{m-1} \succ ... \succ 1$ in the article "On the reducibility of a continuous function..." [Reports of Acad. Sci. USSR, 1961, 130, No. 5], and ended in March 1962 with a submission and acceptance of the aforementioned final article to the journal, where it was published in the beginning of 1964. However, it took another 13 years or more for the rest of the mathematical world to start paying attention to it.

Probably, the first time the words "Sharkovsky ordering" as a mathematical term were used by Peter Kloeden in his article "On Sharkovsky's cycle coexistence ordering" [Bull. Austral. Math. Soc., 1979, vol. 20, 171–177].

Let me now say a few more words about other details directly related to the birth of the Sh-ordering. I was asked many times why it came to my mind to investigate such a topic, not very popular at that time. As I already wrote, I became acquainted with the iterations of functions during the first and second years of study at Taras

© The Author(s), under exclusive license to Springer Nature Switzerland AG 2022
A. M. Blokh and O. M. Sharkovsky, *Sharkovsky Ordering*,
SpringerBriefs in Mathematics, https://doi.org/10.1007/978-3-030-99125-8_5

Shevchenko Kiev State University, where I was involved in mathematical circles[1] and discovered some interesting facts such as, for example, that the iterated sine, $\sin_n(x) = \sin(\sin_{n-1}(x))$, $n = 1, 2, \ldots$, converges to 0 as $\sqrt{3/n}$. And the decision to study one-dimensional iterations came in 1958, at the last, 5th year of study, when it was time to write a graduate thesis ("diploma").

The graduate thesis on iterations was written and successfully defended, but, as it was happening, new questions arose that seemed interesting to the author. So, in November 1958, when I became a graduate student at the Institute of Mathematics of the Academy of Sciences of Ukraine, the problem of choosing the subject of research did not exist for me. However, there was a problem with the choice of supervisor: All potential supervisors insisted that their students should work on projects related to what they themselves were working upon. Eventually, at the suggestion by Yuri Alekseevich Mitropolsky, who was the Director of the Institute, it was approved that the official supervisor will be Nikolai Nikolaevich Bogolyubov, who already had moved to Dubna (near Moscow).

After the first year of graduate school, which was mainly devoted to the preparation for the so-called "candidate exams", I was able to actively engage in research. After a couple of years, in June 1961, my thesis underwent preliminary consideration, and, as a result, was admitted to the official defense. I defended by PhD thesis on October 28, 1961. The thesis was entitled "On some problems of the theory of one-dimensional iterative processes", and it was based on four articles by the author, three of which were published in 1960–61 in UMZh, and the fourth one in the aforementioned Reports of the Academy of Sciences of the USSR. Thus, the Ph.D. thesis already contained a part of the o r d e r i n g $\forall k \neq 2^i \succ \ldots 2^m \succ 2^{m-1} \succ \ldots \succ 1$, and since Reports of the Academy of Sciences of the USSR were translated (by the AMS) into English already at that time, this statement became also available to English-speaking readers.

Yuri Makarovich Berezansky was an official opponent to my thesis. He worked at our institute, and I met him many times at that time discussing my dissertation and other problems. Since my head was busy "clarifying the details" on the coexistence of cycles, I talked to him about my progress in this direction, and Yu. M. expressed his doubts that specific "details" (or parts) of the coexistence ordering for cycles actually occur, because it sounded very unusual. However, soon after the defense that took place on October 28, 1961, within 2 days the proof crystallized out in my mind as a whole, and then, as I remember very well, it took (as many as!) 11 days to put everything on paper. The title arose "by itself ": At that time, the political term "peaceful coexistence of two systems, capitalism and socialism" was used very often in mass media, and it seemed that the word c o e x i s t e n c e had to be highly appropriate for the situation that loomed with periods of cycles (although, perhaps, the word "forcing" would reflect the essence of the matter more accurately).

It took additional 3 months to finalize the draft version, to print the handwritten text on a typewriter in several copies as required by the journal staff, and then to write the formulas in all printed copies by hand (this was the usual procedure).

---

[1] This is how we used to call informal regular gatherings of students interested in mathematics.

The article also included more than ten drawings that had to be made on separate sheets. Finally, in March of 1962, the article was sent to the Ukrainian Mathematical Journal, and the editors sent it for review. As Yu. M. told me later, at his suggestion, the article was sent for review to a well-known topologist (it seems to be Albert Solomonovich Schwartz) who could understand the proposed proof and dispel any doubts that might arise. About a year later, a positive review was received and it contained one recommendation to use the term Λ-scheme instead of Λ-construction. The recommendation was accepted by the author, and the replacement was made. I myself had doubts about Lemmas 1–3, which are rather trivial: Is it worth including them? The reviewer clarified the situation by writing that, of course, it is worth having these lemmas in the text for completeness.

After reviewing and editing the text, the manuscript was sent to the printing house, from where I was soon asked how to handle the $\succ$ badge that was not available in the typography (at that time, for each letter or icon, it was necessary to have a cast made from lead, and all the text was typed by a typesetter from such casts by hand). I answered that the easiest way is probably to lay the letter Y on its side, which was done as a result (though they put it on the wrong side).

In 1967, I had my first travel abroad to Prague, where I participated in a conference on nonlinear oscillations. My report was devoted to one-dimensional difference equations and included, in particular, the theorem on the coexistence of periodic solutions with different periods. The organizers published texts of almost all reports in the Proceedings of the conference, but my report was presented by an abstract only ["Proc. 4th Conf. on Nonlinear Oscillations", Academia Publ. House, Prague, 1968, p. 249]: According to the organizing committee, a strange ordering of natural numbers, based upon the simplest difference equation, can hardly be related to a serious theory of nonlinear oscillations.

At that time, many mathematicians would have a similar attitude towards one-dimensional dynamical systems. For example, Yakov Grigor'evich Sinai wrote in his book "Modern Problems of Ergodic Theory" (Fizmatlit, Moscow, 1995): "About twenty years ago I had the general feeling that the structure of one-dimensional dynamical systems is relatively simple and can be fully understood, and at the same time, the results valid for the one-dimensional case do not have natural multidimensional analogs. The years after this have shown that both of these perceptions were wrong. First, new surprising and unexpected patterns were discovered here, and second, some of them are naturally transferred to the case of any dimension". [Lecture 11 "Sharkovsky order and Feigenbaum universality"].

Finally, in 1975, the appearance of the article "Period three implies chaos" by T. Li and J. Yorke attracted the attention of mathematicians to one-dimensional dynamical systems (and, of course, to the notion of *chaos*) and then the article by P. Štefan "A theorem of Sarkovskii on the existence of periodic orbits of continuous endomorphisms of the real line" [Comm. Math. Phys., 1977, 54, 237–248] literally pulled the Sh-ordering out of non-existence showed that very interesting facts in one-dimensional systems had already been found. We can say that from this point the Sh-ordering began its own life, eventful and independent of the author.

Since the original proof was far from optimal, many people were tempted to suggest their own proofs that would be more or less "normal". As a result, in the late 1970s—early 1980s, several proofs were written thanks to the efforts of a few mathematicians or groups of mathematicians. Since the study of one-dimensional systems seemed very promising, it attracted quite a lot of attention. Soon it was already possible to talk about the emergence of a new direction in dynamical systems called *"Combinatorial dynamics"*. Some summaries and prospects of these studies were considered at a special conference "Thirty years after Sharkovskii's theorem: New perspectives" (Murcia, Spain, 1994) [Proceed. Conf.(eds Alseda L., Balibrea F., Llibre J., Misiurewicz M.), Intern. J. Bifurcation and Chaos **5**(5), 1995, and World Sci. Ser. Nonlinear Sci. B, vol. 8, 1996].

*   *   *   *   *

What was the author doing after the paper on the coexistence of periods came out?

Perhaps it would be appropriate to mention and discuss "new surprising and unexpected patterns" which became the focus of his studies. The paper

The reducibility of a continuous function of a real variable and the structure of the fix points of the corresponding iteration process (Russian), Dokl. Akad. Nauk SSSR **139** (1961) no. 5, 1067–1070 (English) indicates that already in 1961 the author was interested in not only periodic orbits (cycles), arguably the simplest types of orbits, but also the global behavior of the process of iteration. This allowed him to get to the bottom of the majority of problems of topological dynamics that appear in the one-dimensional case. In particular, it was possible to develop the basis of the descriptive theory of chaos by applying the descriptive theory of sets to studying (and characterizing) of chaotic behavior of trajectories.

It is time to pull from the oblivion the author's results from 1960 to 1963 (this was more than 50 years ago), and go over the descriptive theory of chaos developed in them! The first here was the paper *On attracting and attracted sets* Dokl. Akad. Nauk SSSR **160** (1965), 1036–1038 (Russian), containing upper descriptive estimates of the complexity of basin of attraction of various attractors for dynamical systems in arbitrary compact spaces. This article was presented to the Reports of the Academy of Sciences of the USSR in 1964 by the academician P. S. Aleksandrov, one of the creators of the descriptive set theory. The author at that time was not yet 28. In the next papers *A classification of fixed points*, Amer. Math. Soc. Transl. **97** (1970) no. 2, 159–179 (transl. from Ukrain. Mat. Zh. **17** (1965), no. 5, 80–95) and *Behavior of a mapping in the neighborhood of an attracting set*, Amer. Math. Soc. Transl. (2) **97** (1970), 227–258 (transl. from Ukrain. Mat. Zh. **18** (1966), no. 2, 60–83) published in 1965-66 and soon translated into English by the AMS, it was proven that all these upper estimates are realized for dynamical systems on the real line. From the descriptive set theory point of view this means that dynamics on the line can be just as complex (chaotic) as dynamics on any other compact space!

It is worth mentioning here that this direction was not as "lucky" as the other one: If the paper on coexistence of cycles on the line translated into English in 1995 has over 1500 citations, the papers devoted to the other direction studied by the author can hardly boast 100 citations (which is quite understandable as it deals with much more complicated notions that, in addition, are "hidden" behind the titles that were not too revealing).

In the end of 2018, namely, from the 5th until the 11th of December, a rather large workshop and conference I W C T A 2018 took place. The list of participants included a number of well-known American experts in the field of dynamical systems and chaotic dynamics, such as J. Auslander, R. Devaney, J. Yorke, and many others. At the workshop I gave two lectures on the Sh-ordering, and at the conference I gave a lecture DESCRIPTIVE THEORY OF CHAOS. In the latter lecture, various applications of the descriptive set theory to the dynamical systems theory were discussed in great detail. In particular, it was shown that if the topological entropy is positive then the associated dynamical system has the following properties:

(1) it has a lot of different attractors of trajectories, namely, the continuum of attractors;
(2) basins of most attractors have a very complex structure, namely, they are sets of the third class in the terminology of the descriptive theory of sets;
(3) basins of different attractors are very intertwined and cannot be separated from each other by open or closed sets, but only by sets of the second class of complexity, and
(4) in the space of all closed subsets of the state space (with the Hausdorff metric), the set of all attractors is an attractor net (network, grid) whose cells are formed by Cantor sets (whose points are themselves attractors of the dynamical system).

These results are based upon the following papers published in the 1960s.

1. Sharkovsky, A. N.: A classification of fixed points. Amer. Math. Soc. Transl. (2) **97**, 159–179 (1970) (transl. from Ukrain. Mat. Zh. **17**(5), 80–95 (1965)).

2. Sharkovsky, A. N.: Behavior of a mapping in the neighborhood of an attracting set. Amer. Math. Soc. Transl. (2) **97**, 227–258 (1970) (transl. from Ukrain. Mat. Zh. **18**(2) 60–83 (1966)).

3. Sharkovsky, A. N.: Partially ordered system of attracting sets. Soviet Math. Dokl. **7**, 1384–1386 (1966)(transl. from Dokl. Akad. Nauk SSSR **170**(5), 1276–1278 (1966)).

4. Sharkovsky, A. N.: Attractors of trajectories and their basins (in Russian), 320p. Naukova Dumka, Kiev (2013) [the book contains the author's thesis of 1966].

One of the days during the conference was my birthday which I celebrated together with the organizers and all the participants of the conference. I conclude this short chapter with several photos from this conference (Figs. 5.1, 5.2, and 5.3).

A FEW PHOTOS TAKEN DURING THE CELEBRATION OF O. M. SHARKOVSKY'S BIRTHDAY

Fig. 5.1 Prof. Jim Yorke (Maryland)

Fig. 5.2 Prof. V. Kannan (Hyderabad) presents O. M. Sharkovsky with ritual shawl

Fig. 5.3 Prof. V. Kannan (Hyderabad)

Appendix

A.1 The Copy of the First Page of the Paper From 1964

A. M. Blokh and O. M. Sharkovsky, *Sharkovsky Ordering*,
SpringerBriefs in Mathematics, https://doi.org/10.1007/978-3-030-99125-8

1964   УКРАИНСКИЙ МАТЕМАТИЧЕСКИЙ ЖУРНАЛ Т. XVI, № 1
              ИНСТИТУТ МАТЕМАТИКИ

Сосуществование циклов
непрерывного преобразования прямой в себя

А. Н. Шарковский

Всякая непрерывная функция действительного переменного $f(x)$, $-\infty < x < +\infty$, порождает непрерывное преобразование T прямой в себя: $x \to f(x)$. Свойства преобразования T определяются в основном структурой множества неподвижных точек преобразования T.

Напомним, что точку α называют неподвижной точкой порядка k преобразования T, если $T^k \alpha = \alpha$, $T^j \alpha \neq \alpha$, $1 \leqslant j < k$. Точки $T\alpha$, $T^2\alpha$, ..., $T^{k-1}\alpha$ также являются неподвижными порядка k и вместе с точкой α составляют цикл порядка k.

В этой работе исследуется вопрос о зависимости между существованием циклов различных порядков.

Основной результат настоящей работы может быть сформулирован в следующей форме. Рассмотрим множество натуральных чисел, в котором введено отношение: n_1 предшествует n_2 ($n_1 \precsim n_2$), если для всякого непрерывного преобразования прямой в себя существование цикла порядка n_1 влечет за собой существование цикла порядка n_2. Такое отношение, очевидно, обладает свойствами рефлексивности и транзитивности, и, следовательно, множество натуральных чисел с этим отношением есть квазиупорядоченное множество[*]. Ниже доказывается

Т е о р е м а. *Введенное отношение превращает множество натуральных чисел в упорядоченное множество и притом упорядоченное следующим образом*

$$3 \prec 5 \prec 7 \prec 9 \prec 11 \prec \ldots \prec 3\cdot2 \prec 5\cdot2 \prec \ldots \prec 3\cdot2^2 \prec 5\cdot2^2 \prec \ldots \prec$$
$$\prec 2^3 \prec 2^2 \prec 2 \prec 1. \qquad\qquad (*)$$

Терминологией упорядоченных множеств в дальнейшем мы пользоваться не будем. Доказательства теорем по существу опираются только на теорему Больцано—Коши о промежуточном значении.

Из непрерывности преобразования T сразу вытекает, что *если у преобразования T существует цикл порядка $k > 1$, то преобразование T имеет и неподвижную точку первого порядка*.

Т е о р е м а 1. *Если преобразование T имеет цикл порядка $k > 2$, то оно имеет и цикл второго порядка*[**].

Пусть α_1, α_2, ..., α_k — точки цикла, причем $T\alpha_i = \alpha_{i+1}$, $i = 1, 2, \ldots$, $k-1$, $T\alpha_k = \alpha_1$. Пусть $\alpha_i < \alpha_l (i \neq l)$, $\alpha_r > \alpha_i (i \neq r)$. Рассмотрим интервал (α_1, α_{r-1}) (считаем, что $r > 2$; если $r = 2$, следует взять интер-

---

[*] Г. Б и р к г о ф, Теория структур, Гостехиздат, М., 1952, стр. 16—21.
[**] Это утверждение содержится в [1]. Здесь дается более четкое его доказательство.

A.2 The Copy of the Last Page of the Paper From 1964

Теорема 7. *Между любыми двумя точками цикла порядка* $k > 1$ *лежит хотя бы одна точка цикла порядка* $l < k$.

Пусть $\alpha > \beta$ — точки цикла порядка k; n_α, n_β — количество точек этого цикла, меньших соответственно точек α и β. Очевидно, $k > n_\alpha > > n_\beta \geqslant 0$. Существует n_α различных целых положительных чисел s_i, $i = 1, 2, \ldots, n_\alpha$, меньших k и таких, что $T^{s_i}\alpha < \alpha$. Так как $n_\alpha > n_\beta$, найдется s_{i_0}, $1 \leqslant i_0 \leqslant n_\alpha$, такое, что $T^{s_{i_0}}\alpha < \alpha$, $T^{s_{i_0}}\beta > \beta$. А это означает, что существует точка $\gamma \in (\beta, \alpha)$, для которой $T^{s_{i_0}}\gamma = \gamma$; γ есть точка цикла порядка $l \leqslant s_{i_0} < k$.

В заключение отметим еще, что все результаты можно перевести на язык периодических решений функционального уравнения $y(x + 1) = = f(y(x))$ (x пробегает дискретную последовательность значений). Например, если преобразование прямой в себя $y \to f(y)$ непрерывно, то 1) если функциональное уравнение имеет периодическое решение с периодом k, то у него есть и периодические решения с любым периодом, следующим в (∗) за k, 2) если уравнение не имеет периодического решения с периодом k, то у него нет периодических решений ни с каким периодом, предшествующим k в (∗).

Автор приносит благодарность Ю. М. Березанскому и Ю. А. Митропольскому, ознакомившимся с рукописью работы и давшим ряд полезных советов.

ЛИТЕРАТУРА

1. А. Н. Шарковский, УМЖ, т. XII, № 4, 1960.
2. А. Н. Шарковский, ДАН СССР, т. 139, № 5, 1961.

Поступила 22.III 1962 г.
Киев

Co-existence of the cycles of a continuous mapping of the line into itself

A. N. Sharkovsky

Summary

The basic result of this investigation may be formulated as follows. Consider a set of natural numbers in which the following relationship is introduced: n_1 precedes n_2 ($n_1 \preceq n_2$), if for any continuous mappings of the real line into itself the existence of a cycle of order n_2 follows from the existence of a cycle of order n_1. The following theorem holds.

Theorem. The introduced relationship transforms the set of natural numbers into an ordered set, ordered in the following way:
$$3 \prec 5 \prec 7 \prec 9 \prec 11 \prec \ldots \prec 3 \cdot 2 \prec 5 \cdot 2 \prec \ldots \prec 3 \cdot 2^2 \prec 5 \cdot 2^2 \prec \ldots \prec 2^3 \prec 2^2 \prec 2 \prec 1.$$

A.3 Translation of the Original Paper From 1964

The following article was published in 1996. The full data of this publication is as follows:

Coexistence of cycles of a continuous map of the line into itself, A. N. Sharkovskiĭ, Translated from the Russian original by J. Tolosa. Proceedings of the Conference

"Thirty Years after Sharkovskiĭ's Theorem: New Perspectives" (Murcia, 1994). Internat. J. Bifur. Chaos Appl. Sci. Engrg. 5 (1995), no. 5, 1263–1273. ©1996, World Scientific (Singapour).

Coexistence of cycles of a continuous map of the line into itself
A. N. Sharkovsky

Every continuous function of a real variable $f(x)$, $-\infty < x < \infty$, generates a continuous map T of the line into itself: $x \mapsto f(x)$. The properties of the map T are basically determined by the structure of the set of its periodic points.

Recall that a point α is called a periodic point of period k of the map T if $T^k \alpha = \alpha$ and $T^j \alpha \neq \alpha$ for $1 \leq j < k$. The points $T\alpha$, $T^2\alpha$, ..., $T^{k-1}\alpha$ are also periodic points of period k, and together with α they form a cycle of period k.

In this paper we study the problem of the dependence between the existence of cycles of various periods.

The main result of this paper may be stated as follows. Consider the set of natural numbers, in which the following relation has been introduced: n_1 precedes n_2 ($n_1 \preceq n_2$) if for every continuous map of the line into itself the existence of a cycle of period n_1 implies the existence of a cycle of period n_2. This relation is clearly reflexive and transitive and, consequently, the set of natural numbers with this relation is a quasi-ordered set.[1] It is proved below

Theorem *This relation turns the set of natural numbers into an ordered set, which is ordered in the following way:*

$$3 \prec 5 \prec 7 \prec 9 \prec 11 \prec \cdots \prec 3 \cdot 2 \prec 5 \cdot 2 \prec \cdots \prec 3 \cdot 2^2 \prec 5 \cdot 2^2 \prec \cdots$$

$$\cdots \prec 2^3 \prec 2^2 \prec 2 \prec 1 \qquad\qquad (*)$$

The terminology of ordered sets will not be used in the sequel. The proofs of the theorems actually rely only on Bolzano-Cauchy's intermediate value theorem.

The continuity of T immediately implies that if the map T has a cycle of period $k > 1$ then it also has a fixed point.

Theorem A.1 *If the map T has a cycle of period $k > 2$ then it also has a cycle of period two.* [2]

Let $\alpha_1, \alpha_2, ..., \alpha_k$ be the points of the cycle, with $T\alpha_i = \alpha_{i+1}$, $i = 1, 2, ..., k-1$, $T\alpha_k = \alpha_1$. Assume that $\alpha_1 < \alpha_i$ ($i \neq 1$) and $\alpha_r > \alpha_i$ ($i \neq r$). Consider the interval (α_1, α_{r-1}) (we assume that $r > 2$; if $r = 2$ one must take the interval (α_k, α_r)). According to whether (α_1, α_{r-1}) contains fixed points or not, we denote by β either the fixed point closest to α_{r-1} or the point α_1 (if (α_1, α_{r-1}) contains fixed points, the nearest point to α_{r-1} exists by the continuity of T). Since $T\alpha_{r-1} = \alpha_r > \alpha_{r-1}$, then $T x > x$ for $x \in (\beta, \alpha_{r-1}]$. If β is a fixed point then, as can be easily seen, for every

---

[1] G. Birkhoff, *Lattice Theory*, Amer. Math. Soc., New York, 1948.

[2] This assertion is in Sharkovsky [1.]. Here we give a more accurate proof.

integer $j > 0$ there is a neighborhood of β such that $T^j x > x$ for every $x > \beta$ in this neighborhood. If $\beta = \alpha_1$ then $T^j \beta = \alpha_{j+1} > \alpha_1 = \beta$ for $0 < j < k$. On the other hand, $T^{k-r+2}\alpha_{r-1} = \alpha_1 < \alpha_{r-1}$. Therefore, by the continuity of T, there is a point γ on (β, α_{r-1}) such that $T^{k-r+2}\gamma = \gamma$. Since $T\gamma \neq \gamma$, then γ is a periodic point of period l, where $1 < l \leq k - r + 2 < k$. And since there is always a periodic point of period smaller than k, but bigger than one, then there is always a periodic point of period two.

The statements and proofs of the subsequent assertions will be preceded by the following rather trivial lemmas, whose proof is given only for the sake of completeness.

Lemma A.1 *If $T^p \alpha = \alpha$ and α is a periodic point of period k of the map T, then p is a multiple of k.*

Indeed, if α is a periodic point of period k then $T^k \alpha = \alpha$ and $T^j \alpha \neq \alpha$ for $j < k$. Let $p = ks + r$, $r < k$. If we assume that $r \neq 0$ then $T^r \alpha \neq \alpha$ and $T^p \alpha = T^r \underbrace{T^k \ldots T^k}_{s \text{ times}} \alpha \neq \alpha$.

Lemma A.2 *If T has a periodic point α of period $k = 2^n l$, with l odd, then for the map $S = T^{2^m}$ the point α is a periodic point of period*

$$
q = \begin{cases} 2^{n-m}l, & \text{if } n \geq m, \\ l, & \text{if } n \leq m. \end{cases}
$$

Proof By Lemma A.1, $T^p \alpha = \alpha$ only for $p = ki$, $i = 1, 2, \ldots$ Assuming that α is a periodic point of S, let us find its period q. We have $S^q \alpha = \alpha$ and $S^j \alpha \neq \alpha$ for $1 \leq j < q$. Since $S^q = T^{2^m q}$ then $S^q \alpha = \alpha$ if and only if $2^m q = ki$, where i is a natural number. Hence, $q = \frac{k}{2^m} i$. The least value of i for which the right-hand side is an integer corresponds to the desired value of q. Indeed, for this q we have, as one can easily see, $S^q \alpha = \alpha$ and $S^j \alpha \neq \alpha$ when $j < q$.

If $k = 2^n l$, with l odd, then $q = 2^{n-m} l i$. For $n \geq m$ we have $i = 1$ and, therefore, $q = 2^{n-m} l$. For $n < m$ we have $i = 2^{m-n}$, i.e. $q = l$. $\qquad\square$

Corollary *Under the assumptions of Lemma A.2, if $l > 1$ then the periodic point α of the map S has period higher than two.*

Lemma A.3 *A point α is a periodic point of period 2^m of the map T if and only if $T^{2^m} \alpha = \alpha$ and $T^{2^{m-1}} \alpha \neq \alpha$.*

The condition is clearly necessary; let us show it is sufficient. If $T^{2^m} \alpha = \alpha$ then α may be a periodic point of period 2^j, $j = 0, 1, \ldots, m$ (Lemma A.1). Since $T^{2^{m-1}} \alpha \neq \alpha$ then we also have $T^{2^j} \alpha \neq \alpha$ for every $j < m - 1$, since $T^{2^{m-1}} = \underbrace{T^{2^j}(T^{2^j}(\ldots T^{2^j})\ldots)}_{2^{m-j-i} \text{ times}}$.

Thus, α is a periodic point of period 2^m.

Theorem A.2 *If the map T has a cycle of period 2^n, $n > 1$, then T has cycles of period 2^i for every $i = 1, 2, ..., n - 1$.*[3]

Let α be a periodic point of period 2^n. We show that T has a periodic point of period 2^m for $1 \leq m < n$.

Set $T^{2^{m-1}} = S$. By Lemma A.2, α is a periodic point for S of period 2^{n-m+1}, i.e., of period higher than two. By Theorem A.1, S has a periodic point β of period two: $S^2\beta = \beta$ and $S\beta \neq \beta$. Consequently, $T^{2^m}\beta = \beta$ and $T^{2^{m-1}}\beta \neq \beta$.

The following theorem is proved analogously.

Theorem A.3 *If the map T has a cycle of period k and k is not a power of two then T has cycles of periods 2^i for $i = 1, 2, 3, ...$*

Let α be a periodic point of period k. We show that T has a periodic point β of period 2^m, where $m \geq 1$.

Set $T^{2^{m-1}} = S$. By the corollary to Lemma A.2, α is a periodic point of period higher than two for S. By Theorem A.1, S has a periodic point β of period 2. Thus, $S^2\beta = \beta$ and $S\beta \neq \beta$, i.e., $T^{2^m}\beta = \beta$ and $T^{2^{m-1}}\beta \neq \beta$.

From Theorem A.3 it follows that there are maps having cycles of arbitrarily high period, since it is always easy to construct a map having a cycle of a prescribed period – in particular, an period different from a power of two.

Theorem A.3 also shows that it suffices to specify the function $f(x)$, defining the map T, at finitely many points (forming a cycle), for example, at three points, and there will exist infinitely many cycles, independently of the way we (continuously) change the values of $f(x)$ at the remaining points of the line.

Let us consider the set of periodic points in one cycle. Assume that the points α_1, $\alpha_2 = T\alpha_1, ..., \alpha_k = T\alpha_{k-1}$ form a cycle of period k. Let us divide the points of the cycle into two sets M_1 and M_2 so that $\alpha_i \in M_1$ if $\alpha_i < T\alpha_i$ and $\alpha_i \in M_2$ if $\alpha_i > T\alpha_i$. Let $\alpha^{M_1} = \max_{\alpha_i \in M_1}$ and $\alpha^{M_2} = \min_{\alpha_i \in M_2}$. We have two possibilities: either $\alpha^{M_1} < \alpha_{M_2}$ or $\alpha^{M_1} > \alpha_{M_2}$.

Lemma A.4 *If $\alpha^{M_1} > \alpha_{M_2}$ then the map T has cycles of any period.*

Among all the points belonging to M_1 and bigger than M_2 let us chose one at which the value of the function $f(x)$ defining the map T is the largest. Denote it by β. Since $T\alpha_{M_2} < \alpha_{M_2}$ and $T\beta > \beta$, the set of all periodic points on the interval (α_{M_2}, β) is nonempty and closed (by the continuity of T). Let γ be the greatest fixed point on this interval. Then $T\gamma = \gamma$ and $Tx > x$ on $(\gamma, \beta]$. The interval $(\gamma, T\beta]$ has been chosen so that it contains periodic points of the given cycle (for example, β and $T\beta$). Since applying T repeatedly to any point of the cycle one must close the cycle, then $(\gamma, T\beta]$ must contain at least one point δ of the cycle such that either $T\delta > T\beta$ or $T\delta < \gamma$. The first inequality is impossible. Indeed, if $\delta \in M_2$ then $T\delta < \delta < T\beta$, and if $\delta \in M_1$ then $T\delta < T\beta$ by our choice of β. Thus, on $(\gamma, T\beta]$ there is a point δ of the cycle for which $T\delta < \gamma$ (it is possible that $\delta = T\beta$.

---

[3] The statements of Theorems A.2 and A.3 are in Sharkovsky [2].

Fig. A.1 An L-scheme

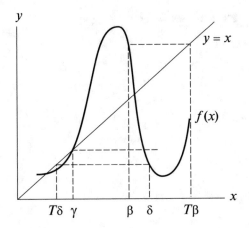

Since $Tx > x$ on $(\gamma, \beta]$ then $\delta \in (\beta, T\beta]$. The scheme thus obtained (Fig. A.1): $T\gamma = \gamma$, $Tx > x$ on $(\gamma, \beta]$, $\delta \in (\beta, T\beta]$, and $T\delta < \gamma$ (we shall call it an *L-scheme*) guarantees the existence of cycles of all periods.

Indeed, $T(\gamma, \beta] \supseteq (\gamma, T\beta]^4$ and, consequently, $(\gamma, \beta]$ contains a closed nonempty set of points that is mapped by T, in one step, into the point δ. Denote by δ_1 the smallest of these points. Analogously, since $T(\gamma, \delta_1] = (\gamma, \delta]$, then $(\gamma, \delta_1]$ contains a nonempty closed set of points that are carried by T, in one step, into δ_1. Denote by δ_2 the smallest of these points. Clearly, $\gamma < \delta_2 < \delta_1$ and $T(\gamma, \delta_2] = (\gamma, \delta_1]$. Continuing the process of construction of the points δ_i, we obtain a sequence $\delta > \delta_1 > \delta_2 > \dots > \delta_{i-1} > \delta_i > \dots > \gamma$, such that $T\delta_i = \delta_{i-1}$. Evidently, $T^i \delta_{i-1} = T\delta$ and $T^i \delta_i = \delta$. Thus, $T^i \delta_i > \delta_i$ and $T^i \delta_{i-1} < \delta_{i-1}$, so that by the continuity of T^i there is at least one point ρ_i on (δ_i, δ_{i-1}) such that $T^i \rho_i = \rho_i$. Since $T^j (\gamma, \delta_{i-1}] = (\gamma, \delta_{i-j-1}] \subset (\gamma, \delta_1]$ for $j < i - 1$ and $Tx > x$ on $(\gamma, \delta_1]$, then $T^j x > x$ on $(\gamma, \delta_{i-1}]$ for $1 \leq j < i$. Hence, $T^j \rho_i \neq \rho_i$ when $1 \leq j < i$, i.e., ρ_i is a periodic point of period i.

This finishes the proof of the lemma.

Remark If there is a fixed point that is less than α^{M_1} (but greater than $\alpha_{min} = \min_{i=1,2,\dots,k} \alpha_i$) then, as before, the map T contains an L-scheme. Hence it follows that, independently of the distribution of the points of the cycle, T has cycles of any period.

If there is a fixed point that is greater than α_{M_2} (but smaller than $\alpha_{max} = \max_{i=1,2,\dots,k} \alpha_i$) then T has a scheme representing the reflection of an L-scheme with respect to the point γ as a center. As in the proof of Lemma A.4, one shows that this scheme also guarantees the existence of cycles of all periods.

Let us consider the case when $\alpha^{M_1} < \alpha_{M_2}$. We have the following result.

Lemma A.5 *If $\alpha^{M_1} < \alpha_{M_2}$ and there is a point $\alpha \in M_1$ such that $T\alpha \in M_1$ as well, then the map T has cycles of odd periods greater than k and cycles of all even periods.*

---

[4] By $T(\gamma, \beta]$ we mean the set of images of points belonging to $(\gamma, \beta]$.

Fig. A.2 An M-scheme

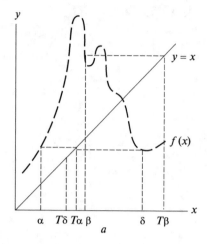

Fig. A.3 Dynamics on the interval $[\beta, \gamma]$

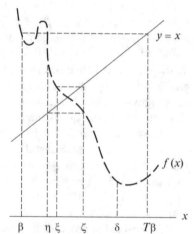

The lemma also holds for $\alpha \in M_2$ and $T\alpha \in M_2$.

Let us start by singling out a scheme that will lead to the proof of the lemma.

Let $n = \min\{j : T^j\xi \leq \alpha, \xi \in M_1, \xi \geq T\alpha\}$ and let β be the ξ at which this minimum is attained (or one of them, if there are several such points). Thus, $T^n\beta = \gamma \leq \alpha$ and $T^i\beta > \alpha$ for $i < n$. Let us consider the sequence $T\beta, T^2\beta, T^3\beta, \ldots$ Let $T^l\beta$ be the first point belonging to M_1. It is easy to see that $T^l\beta < T\alpha$. Indeed, if we had $T^l\beta > T\alpha$ (it is clear that $T^l\beta \neq T\alpha$) then the min j indicated earlier would be less than n. Since $T\beta > \beta \geq T\alpha$ then $T\beta \in M_2$ and, consequently, $l \geq 2$. Denote the point $T^{l-1}\beta$ by δ. Then $\delta \in (\beta, T\beta]$ and $T\delta < T\alpha$. This discussion leads to the picture shown in Fig. A.2. We shall call such a scheme, an *M-scheme*.

Consider the interval $[\beta, \delta]$ (Fig. A.3). Let η be the greatest of the points x of this interval for which $Tx = T\beta$ (it is possible that $\eta = \beta$). On (η, δ) we have $Tx < T\beta$. Since $T\eta = T\beta \geq \delta$ and $T\delta < T\alpha \leq \eta$ then on (η, δ) there is at least one point ζ

Fig. A.4 An approximate
graph of the function
$f(f(x))$

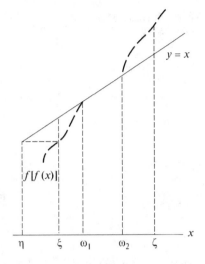

such that $T\zeta = \eta$. If there are several such points on (η, δ), we shall assume that
ζ is the smallest of them. Thus, $T\eta = T\beta$, $T\zeta = \eta$, and $\eta < Tx < T\beta$ for every
$x \in (\eta, \zeta)$. Further, let us choose the point ξ in $[\eta, \zeta]$ that is the largest of all x for
which $Tx = \zeta$. For every $x \in (\xi, \zeta)$ we have $\eta < Tx < \zeta$.

In order to better illustrate the subsequent arguments, let us construct an approximate graph of the function $f(f(x))$ on (ξ, ζ) (Fig. A.4).

We have $T^2\xi = \eta < \xi$, $T^2\zeta = T\beta > \zeta$, and $\eta < T^2x < T\beta$ on (ξ, ζ). Let $\omega_1 \le$
ω_2 be respectively the smallest and the largest of the points of (ξ, ζ) for which
$T^2x = x$. Clearly, $T\omega_1 = \omega_2$ and $T\omega_2 = \omega_1$, that is, ω_1 and ω_2 form a cycle of period
two, or, if $\omega_1 = \omega_2 = \omega$, then ω is a fixed point.[5] Moreover, $T(\xi, \omega_1) = (\omega_2, \zeta)$ and
$T(\omega_2, \xi) = (\eta, \omega_1)$.

As we did earlier for the L-scheme, let us construct a sequence $\zeta = \theta_0 >$
$\theta_1 > \theta_2 > \ldots > \omega_2$ such that $T^2\theta_i = \theta_{i-1}$, $T^2(\omega_2, \theta_i) = (\omega_2, \theta_{i-1})$, and a sequence
$\xi = \kappa_0 < \kappa_1 < \kappa_2 \ldots < \omega_1$ such that $T^2\kappa_i = \kappa_{i-1}$ and $T^2(\kappa_i, \omega_1) = (\kappa_{i-1}, \omega_1)$. Consequently, and $T^{2i+1}(\omega_2, \theta_i) = (\eta, \omega_1)$ and $T^{2i+2}(\kappa_i, \omega_1) = (\eta, \omega_1)$ (Fig. A.5).

Since $T\alpha < \eta$ and $T\eta = T\beta > \zeta$, on the interval (α, η) there are points at which
the value of $f(x)$ is equal to $\omega_1, \omega_2, \eta, \theta_i$, and κ_i ($i = 0, 1, 2, \ldots$). We can always
find points $\lambda_1, \lambda_2, \mu_0, \nu_{-1} \in (\alpha, \eta)$, such that $T\lambda_1 = \omega_1$, $T\lambda_2 = \omega_2$, $T\mu_0 = \theta_0 =$
ζ, $T\nu_{-1} = \eta$ and $T(\nu_{-1}, \lambda_1) = (\eta, \omega_1)$, $T(\lambda_2, \mu_0) = (\omega_2, \zeta)$. Further, we can find
points μ_i, $i = 1, 2, \ldots$, such that $T\mu_i = \theta_i$, $T(\lambda_2, \mu_i) = (\omega_2, \theta_i)$, and points ν_i,
$i = 0, 1, 2, \ldots$, such that $T\nu_i = \kappa_i$, $T(\nu_i, \lambda_1) = (\kappa_i, \omega_1)$. Clearly, $T^{2i+2}\mu_i = \eta$,
$T^{2i+2}(\lambda_2, \mu_i) = (\eta, \omega_1)$ and $T^{2i+3}\nu_i = \eta$, $T^{2i+3}(\nu_i, \lambda_1) = (\eta, \omega_1)$.

Since $T\eta = T\beta$, we have $T^n\eta = \gamma$ (n is the least positive integer such that $T^n\beta \le$
α). To pass from one point to the cycle to another we need no more than $k - 1$

---

[5] Indeed, $T\omega_2 \ge \omega_1$, since $T^2(T\omega_2) = T\omega_2$ and, hence, $T\omega_2 \notin (\xi, \omega_1)$. Analogously, $T\omega_1 \le \omega_2$.
Therefore $T[\xi, \omega_1] \supseteq [\omega_2, \zeta]$ and on $[\xi, \omega_1]$ there is a point χ such that $T\chi = \omega_2$. For every
$x \in [\xi, \omega_1]$ we have $T^2x \le x$. Thus, $\chi \ge T^2\chi = T\omega_2 \ge \chi$, whence $T^2\chi = \chi$, i.e., $\chi = \omega_1$.

Fig. A.5 Action of odd and
even iterations of T

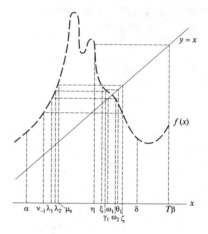

steps, and therefore $n \leq k - 1$. It is not hard to see that if $\gamma = \alpha$ and $\beta = T\alpha$ then $n = k - 1$.

Let us show that the map T has periodic points of odd period greater than k. Let n be even. In this case, $n + 2i + 3$, $i \geq 0$, is odd and there is a periodic point of period $s = n + 2i + 3$. Indeed, $T^s\lambda_1 = \omega_1 > \lambda_1$, $T^s v_i = \gamma < v_i$ and, consequently, on (v_i, λ_1) there are points x such that $T^s x = x$. Let ρ_s be the largest of these points. We claim that ρ_s is a periodic point of period s. Since s is odd, ρ_s can only be a periodic point of odd period (Lemma A.1). Assume that ρ_s is a periodic point of period r, where $r < s$ is odd. We have $T\rho_s \in (\kappa_i, \omega_1)$, and there is a point $\pi' \in (\kappa_{i+\frac{s-r}{2}}, \omega_1)$ such that $T^{s-r}\pi' = T\rho_s$. Since we have $T^2 x < x$ on (κ_j, ω_1), $j = 0, 1, 2, \ldots$, and $T^{s-r} = \underbrace{T^2(T^2(\ldots T^2)}_{(s-r)/2}\ldots)$, then $T\rho_s < \pi' < \omega_1$. There is a point π'' such that $\rho_s < \pi'' < \lambda_1$ and $T\pi'' = \pi'$. Thus, $\rho_s < \pi'' < \lambda_1$ and $T^s\pi'' = T^{r-1}T^{s-r}T\pi'' = T^{r-1}T^{s-r}\pi' = T^{r-1}T\rho_s = \rho_s < \pi''$ and, therefore, on (π'', λ_1) there is a point (π'', λ_1) at which $T^s\rho_s' = \rho_s'$; but $\rho_s < \rho_s'$, which contradicts the fact that ρ_s is the largest of the points $x \in (v_i, \lambda_1)$ such that $T^s x = x$.. The odd number $s = n + 3(i = 0)$ is never bigger than the smallest odd number bigger than k and, therefore, for n even, we have proved the existence of periodic points of odd period bigger than k.

If n were odd, one would have to use the sequence of points μ_i, instead of the sequence $\{v_i\}$.

Now we prove that the map T has periodic points of arbitrary even period. Let n be even. In this case one must use the sequence $\{\mu_i\}$. Set $s = n + 2i + 2$; then $T^s\lambda_2 = \omega_1 > \lambda_2$, $T^s\mu_i = \gamma < \mu_i$ and, therefore, on (λ_2, μ_i) there are points x such that $T^s x = x$. Let σ_s be one such point. We claim that for $s \geq 2k - 2$, σ_s is a periodic point of period s. Indeed, since $T^s\sigma_s = \sigma_s$, then σ_s is either a periodic point of period s or a periodic point of smaller period r, and s is a multiple of r (Lemma A.1). Clearly, $r \leq s/2$ and hence if $T^j\sigma_s \neq \sigma_s$ for $1 \leq j \leq s/2$ then σ_s is a periodic point of period s. On (λ_2, μ_i) we have $T^j x > \eta > x$ for every

$1 \le j \le s - n$, since $T^j(\lambda_2, \mu_i) \subset (\eta, \zeta)$ for $j < s - n$. Thus, σ_s is a periodic point of period s for $s - n \ge s/2$; and this inequality is always true for $s \ge 2k - 2$.

When n is odd, the existence of periodic points of even period $s \ge 2k - 2$ is proved analogously, only now using the points v_i.

It remains to prove that T has periodic points of even period smaller than $2k - 2$. Before completing the proof of Lemma A.5, let us prove the following result.

Lemma A.6 *If the map T has a cycle of odd period then it has cycles of any even period.*

Consider the sets M_1 and M_2. If $\alpha^{M_1} > \alpha_{M_2}$ then there are cycles of all periods (Lemma A.4). Assume that $\alpha^{M_1} < \alpha_{M_2}$. The points of a cycle of odd period k will also form a cycle of period k for the map $S = T^2$ (Lemma A.2). For the map S one can construct sets M_1^2 and M_2^2, as we did with the sets M_1 and M_2, considering that a point α_i is in M_1^2 if $\alpha_i < T^2\alpha_i$ and $\alpha_i \in M_2^2$ if $\alpha_i > T^2\alpha_i$. Let $\alpha^{M_1^2}$ be the largest point of M_1^2 and let $\alpha_{M_2^2}$ be the smallest point in M_2^2. Let us prove that S has cycles of all periods.

Since $\alpha^{M_1} < \alpha_{M_2}$, the map T has a fixed point γ such that $\alpha^{M_1} < \gamma < \alpha_{M_2}$. This is a fixed point also for $S = T^2$. If $\alpha^{M_1^2} \ne \alpha^{M_1}$ (and, consequently, also $\alpha_{M_2^2} \ne \alpha_{M_2}$) then either $\alpha^{M_1^2} \in M_2$ and $\gamma < \alpha^{M_1^2}$ or $\alpha_{M_2^2} \in M_1$ and $\gamma > \alpha_{M_2^2}$. It remains to use the remark following Lemma A.4.

Assume that $\alpha^{M_1^2} = \alpha^{M_1}$ and hence also $\alpha_{M_2^2} = \alpha_{M_2}$, $M_1^2 = M_1$, and $M_2^2 = M_2$. Let α_1 be the smallest of all the α_i, $i = 1, 2, \ldots, k$ (Figs. A.6 and A.7).

Then $\alpha_k \in M_2$.. Since $\alpha_{k-1} > \alpha_1$ then $\alpha_{k-1} \in M_2^2$ and, consequently, $\alpha_{k-1} \in M_2$. Thus, $\alpha_{k-1} > \alpha_k$.. Let α_r be the largest of all points α_i, $i = 1, 2, \ldots, k$. Then $\alpha_{r-1} \in M_1$ and hence $\alpha_1 < \alpha_{r-1} < \alpha_k$. Since on (α_k, α_{k-1}) the function $f(x)$ takes at least all the values of the interval (α_1, α_k), then on (α_k, α_{k-1}) there is a point δ such that $T\delta = \alpha_{r-1}$. Finally, let ω be the largest of the points $x \in [\gamma, \delta]$ for which $Sx = x$ (there is at least one point x such that $Sx = x$, since $S\gamma = \gamma$).

Fig. A.6 The first figure illustrating the proof of Lemma A.6

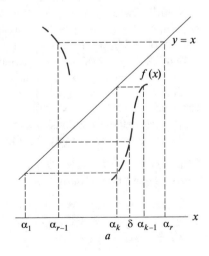

Fig. A.7 The second figure
illustrating the proof of
Lemma A.6

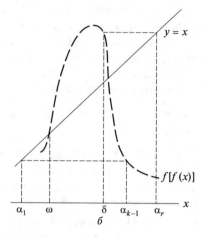

Thus, we have: $S\omega = \omega$, $S\delta = \alpha_r > \delta$, $Sx > x$ on $(\omega, \delta]$, $\alpha_{k-1} \in (\delta, \alpha_2)$, and $S\alpha_{k-1} = \alpha_1 < \omega$ (Fig. A.6). We have singled out an L-scheme for the map S (see the proof of Lemma A.4), which guarantees the existence of cycles of all periods for S.

The fact that S has cycles of all periods immediately implies the existence of cycles of even period for T. Let us prove, for example, that T has a cycle of period $l = 2l_1$.

Let α be a periodic point of period l_1 of S. This means that $S^{l_1}\alpha = \alpha$ and $S^{j_1}\alpha \neq \alpha$ for $1 \leq j_1 < l_1$, i.e., $T^l\alpha = \alpha$ and $T^j\alpha \neq \alpha$, where j is any even number less than l. Since $S\alpha \neq \alpha S$ then also $T\alpha \neq \alpha$. Hence, either α is a periodic point of period l for T or α is a periodic point of odd period $l_2 > 1$, and $l_2 < l_1$. But to a cycle of odd period we can always apply either Lemma A.4 or Lemma A.5. Indeed, since the cycle contains an odd number of points, then there are more points either in M_1 or in M_2. To fix the ideas, assume that there are more points in M_1 than in M_2. Then necessarily there is a point $\mu \in M_1$ such that $T\mu \in M_1$, since otherwise the number of points of M_1 could not be bigger than that of M_2. Thus, since the map T, having a cycle of period l_2, satisfies the assumptions of either Lemma A.4 or Lemma A.5, then T must have cycles of any even period $\geq 2l_2 - 2$; and hence, of period l (because $\geq 2l_2 - 2$). Lemma A.6 is proved.

This concludes the proof of Lemma A.5. Since in its proof we have already established the existence of cycles of odd period (bigger than k), it follows that there are cycles of all even periods as well.

All these arguments imply the following result.

Theorem A.4 *If the map T has a cycle of odd period k then it has cycles of all odd periods bigger than k and all even periods.*

Theorem A.4 cannot be sharpened. Now we will construct an example of a map T having a cycle of period $2m + 1$ but having no cycles of period $2j - 1$ for $j = 2, 3, ..., m$.

Fig. A.8 A map that shows that Theorem A.4 is sharp

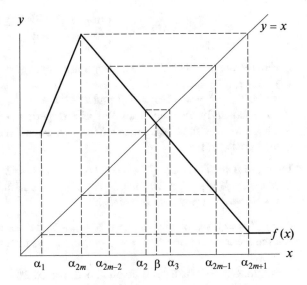

Assume that the points α_i, $i = 1, 2, \ldots, 2m + 1$, form a cycle of period $2m + 1$, with $\alpha_{i+1} = T\alpha_i$, $i = 1, 2, \ldots, 2m$, $\alpha_1 = T\alpha_{2m+1}$, and assume that $\alpha_1 < \alpha_{2m} < \alpha_{2m-2} < \ldots < \alpha_2 < \alpha_3 < \alpha_{2m+1}$. Assume that the continuous function $f(x)$ defining the map T is equal to α_2 for $x \leq \alpha_1$, is equal to α_1 for $x \geq \alpha_{2m+1}$, and for $\alpha_1 \leq x \leq \alpha_{2m+1}$ is a piecewise-linear function with vertices at the points (α_1, α_2), (α_2, α_3), ..., $(\alpha_{2m}, \alpha_{2m+1})$, and $(\alpha_{2m+1}, \alpha_1)$ of the x, y-plane (Fig. A.8).

It is not hard to see that

$$T^{2j-1}(\alpha_1, \alpha_{2m}] = (\alpha_{2j}, \alpha_{2m+1}],$$

$$T^{2j-1}(\alpha_{2i+2}, \alpha_{2i}] = \begin{cases} [\alpha_{2(i+j)-1}, \alpha_{2(i+j)+1}), & \text{if } 2 \leq i + j \leq m, \\ (\alpha_{2(i+j-m)}, \alpha_{2m+1}], & \text{if } m + 1 \leq i + j \leq 2m - 1, \end{cases}$$

$$T^{2j-1}(\alpha_{2i+1}, \alpha_{2i+3}] = \begin{cases} (\alpha_{2(i+j)+2}, \alpha_{2(i+j)}], & \text{if } 2 \leq i + j \leq m - 1, \\ (\alpha_1, \alpha_{2m+1}], & \text{if } i + j = m, \\ [\alpha_1, \alpha_{2(i+j-m)+1}), & \text{if } m + 1 \leq i + j \leq 2m - 1, \end{cases}$$

$$i = 1, 2, \ldots, m - 1, \quad j = 1, 2, \ldots, m.$$

If $T\beta = \beta$, then

$$T^{2j-1}(\alpha_2, \beta) = (\beta, \alpha_{2j+1}], \quad 1 \leq j \leq m,$$

$$T^{2j-1}(\beta, \alpha_3) = \begin{cases} (\alpha_{2j+2}, \beta), & \text{if } 1 \le j < m, \\ (\alpha_1, \beta), & \text{if } j = m. \end{cases}$$

Finally, observe that $T^{2j-1}x = \alpha_{2j}$ for every $x \le \alpha_1$ and $T^{2j-1}x = \alpha_{2j-1}$ for $x \ge \alpha_{2m+1}$, $j = 1, 2, \ldots, m$..

Thus, $T^{2j-1}x > x$ when $x < \beta$ and $T^{2j-1}x < x$ if $x > \beta$, for every $1 \le j \le m$, and, consequently, the map T has no cycles of period $3, 5, \ldots, 2m - 1$.

Theorem A.4 can be generalized to the case when T has a cycle of any period that is not a power of two.

Theorem A.5 *If the map T has a cycle of period $k = 2^n l$, where $l > 1$ is odd, then T has a cycle of period $2^n r$, where $r > l$ is any odd number, and cycles of period $2^{n+1}s$, where s is any natural number.*

Proof If $n = 0$ we obtain Theorem A.4, which has already been proved. Assume that the theorem is true for $n = m - 1$, and let us then prove that it also holds for $n = m$.

Assume that T has a periodic point α of period $2^m l$. Let us prove, for instance, that in this case T also has a periodic point of period $2^m r_o$, where $r_0 > l$ is odd. The point α is a periodic point of period $2^{m-1} l$ for the map $S = T^2$ (Lemma A.2) and, by our assumption, s must have a periodic point β of period $2^{m-1} r_0$. This means that $S^{2^{m-1}r_0}\beta = \beta$ and $S^j\beta \ne \beta$ for $j = 1, 2, 3, \ldots, 2^m r_0 - 1$, that is, $T^{2^m r_0}\beta = \beta$ and $T^i\beta \ne \beta$ for every even i less than $2^m r_0$. We have $T\beta \ne \beta$, since otherwise we would have $S\beta = \beta$. Thus, either β is a periodic point of period $2^m r_0$ for T, or β is a periodic point of odd period, and then, by Theorem A.3, T has periodic points of every even period, and, therefore, there is a periodic point γ of period $2^m r_0$.

The proof that T has also periodic points of period $2^{m+1}s$, where s is any natural number, is completely similar.

Thus, Theorem A.5 holds for every n.

Theorems A.2, A.3, and A.5, and the fact that there is always a fixed point if there are periodic points of higher period, can be put together in one single theorem. $\square$

Theorem A.6 *If the map T has a cycle of period 2^n, $n > 0$, then T also has cycles of period 2^i, $i = 0, 1, \ldots, n - 1$. If T has a cycle of period $2^n(2m + 1)$, $n \ge 0$, $m > 0$, then it also has cycles of period 2^i, $i = 0, 1, \ldots, n$, and cycles of period $2^n(2r + 1)$, $r = m + 1, m + 2, \ldots$, and of period $2^{n+1}s$, $s = 1, 2, 3, \ldots$.*

Remark Let $\alpha_1, \alpha_2, \ldots, \alpha_k$ be the points of a given cycle of period k, and let $a = min_i\alpha_i$, $b = max_i\alpha_i$. Theorem A.6 concerns only the points of the interval $[a, b]$. Outside $[a, b]$ the map may not have any points of cycles. So, the points of the cycles of the map $\bar{T}$ defined as $\bar{T}x = Ta$ for $x \le a$, $\bar{T}x = Tx$ for $a \le x \le b$, and $\bar{T}x = Tb$ for $x \ge b$, belong to $[a, b]$.

Let us define the *diameter* of the cycle $\alpha_1, \alpha_2, \ldots, \alpha_k$ as the number $d_{\alpha_1, \alpha_2, \ldots, \alpha_k} = max_{1 \le i, j \ge k} |\alpha_i - \alpha_j|$. For every n, following k in (*), there is a cycle $\beta_1, \ldots, \beta_k$ for which $d_{\beta_1, \beta_2 \ldots, \beta_k} < d_{\alpha_1, \alpha_2, \ldots, \alpha_k}$. Moreover, as is easily seen, there is a constant C, depending on $\alpha_1, \alpha_2, \ldots, \alpha_k$, such that for every $m > 1$ following k in (*) there is a cycle $\gamma_1, \ldots, \gamma_m$ for which $d_{\gamma_1, \ldots, \gamma_n} > C$.

Let us construct an example showing that Theorem A.6 completely solves the problem on the existence of cycles of some periods depending on the existence of cycles of other periods.

In the x, y-plane let there be given points $A^{(l)}(x^{(l)}, y^{(l)})$, $A^{(2)}(x^{(2)}, y^{(2)})$, ..., $A^{(k)}(x^{(k)}, y^{(k)})$, with $x^{(1)} < x^{(2)} < ... < x^{(k)}$. These points define the following continuous function $f(x)$: for $x \in [x^{(l)}, x^{(k)}]$, $f(x)$ is a piecewise-linear function with vertices at the points $A^{(1)}, ..., A^{(k)}$; for $x \leq x^{(1)}$, $f(x) = y^{(1)} = const.$, and for $x \geq x^{(k)}$, $f(x) = y^{(k)} = const.$ We shall denote by $T_{A^{(1)} A^{(2)} ... A^{(k)}}$ the map given by this function.

We shall carry out the construction without getting into the details.

Let us take in the plane's two points A_1 and A_2, symmetric with respect to the bisector of the first and third quadrants. It is easy to see that the map $T_{A_1 A_2}$ has only cycles of first and second periods. Let us draw through A_1 a line a_1 perpendicular to the bisector, and through A_2, a line a_2 parallel to the bisector. Let us take on a_1 points A_{11} and A_{12} symmetric with respect to A_1, and on a_2, points A_{21} and A_{22} symmetric with respect to A_2, and such that $|x_{11} - x_{12}| = |x_{21} - x_{22}| \leq \frac{x_1 - x_2}{2}$ (we denote by x_r the x-coordinate of A_r). It can be seen that the map $T_{A_{11} A_{12} A_{21} A_{22}}$ has only cycles of first, second, and fourth periods. Now through the points $A_{11} A_{12}$ and A_{21} we must draw lines a_{11}, a_{12}, and a_{21} perpendicular to the bisector, and through A_{22}, a line a_{22} parallel to the bisector (clearly, a_{11} and a_{12} will coincide with a_1, and a_{22}, with a_2). Next, as before, on these lines one must take points $A_{111}, A_{112}, A_{121}, ..., A_{222}$ symmetric with respect to $A_{11} A_{12}$, $A_{21} A_{22}$ such that (see Figs. A.9 and A.10)

$$|x_{111} - x_{112}| = |x_{121} - x_{122}| = |x_{211} - x_{212}| = |x_{221} - x_{222}| \leq \frac{x_{11} - x_{12}}{2},$$

etc. Observe that one can draw parallel (and perpendicular) lines to the bisector through any points and through any number of these points, provided this number is odd.

Fig. A.9 Dynamics of points $A_{111}, ..., A_{222}$ and $A_{11}, ..., A_{22}$

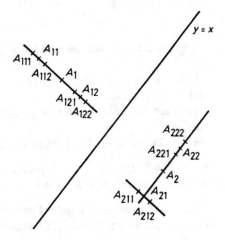

Fig. A.10 Dynamics of
points A_1, A_2 and
$A_{11}, \ldots, A_{22}$

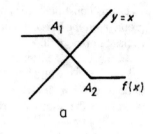

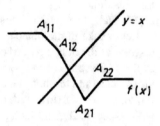

The   map $T_{\underbrace{A_{11\ldots 11}}_{n+1} \underbrace{A_{11\ldots 12}}_{n+1} \cdots \underbrace{A_{22\ldots 22}}_{n+1}}$ has   only   cycles   of   peri-

ods $1, 2, 2^2, \ldots, 2^n + 1$.   To   fix   the   ideas,   assume   that   the   line
$\underbrace{a_{11\ldots 11}}_{n+1}$ is   perpendicular   to   the   bisector.   We   replace   the   two   points

$\underbrace{A_{111\ldots 11}}_{n+1}(x_{11\ldots 11}, y_{11\ldots 11})$, $\underbrace{A_{111\ldots 12}}_{n+1}(x_{11\ldots 12}, y_{11\ldots 12})$ in   the x, y-plane   by   the   points
"from Fig. A.8" $A_{10}(x_{10}, y_{10})$, $A_{20}(x_{20}, y_{20})$, $\ldots$, $A_{2m+1,0}(x_{2m+1,0}, y_{2m+1,0})$ where

$$x_{10} = \underbrace{x_{11\ldots 11}}_{n+1} < x_{2m,0} < x_{2m-2,0} < \cdots < x_{20} < x_{30} < \cdots < x_{2m+1,0} = \underbrace{x_{11\ldots 12}}_{n+1},$$

$$y_{i0} = x_{i+1,0} + (\underbrace{y_{11\ldots 12}}_{n+1} - \underbrace{x_{11\ldots 11}}_{n+1}), \quad i = 1, 2 \ldots, 2m, \quad y_{2m+1,0} = \underbrace{y_{11\ldots 12}}_{n+1}.$$

It is not hard to see that the map $T_{A_{10}A_{20}\cdots A_{2m+1,0}\underbrace{A_{11\ldots 21}}_{n+1} \underbrace{A_{11\ldots 22}}_{n+1}\cdots \underbrace{A_{22\ldots 22}}_{n+1}}$ has cycles of

periods $1, 2, 2^2, \ldots, 2^n, 2^n(2r + 1)$, for $r \geq m$, and $2^{n+1}s$, $s > 0$, and has no cycles
of any other periods.

Theorem A.6 and this example prove the theorem stated at the beginning of this
paper.

The following result is related to Theorems A.1–A.6.

Theorem A.7 *Between any two points of a cycle of period $k > 1$ there is at least
one point of a cycle of period $l < k$.*

Let $\alpha > \beta$ be points of a cycle of period k, and let n_α, n_β be the number of points of
this cycle smaller than α and β, respectively. Clearly, $k > n_\alpha > n_\beta > 0$. There are n_α

distinct positive integers s_i, $i = 1, 2, ..., n_\alpha$, smaller than k and such that $T^{S_i}\alpha < \alpha$. Since $n_\alpha > n_\beta$, there is an s_{i_0}, $1 \leq i_0 \leq n_\alpha$, such that $T^{S_{i_0}}\alpha < \alpha$ and $T^{S_{i_0}}\beta > \beta$. But this means that there is a point $\gamma \in (\beta, \alpha)$ for which $T^{S_{i_0}}\gamma < \gamma$; γ is a point of a cycle of period $1 \leq s_{i_0} \leq k$.

In conclusion, we observe that all the results may be translated into the language of periodic solutions of the functional equation $y(x + 1) = f(y(x))$ (where x runs through a discrete sequence of values). For example, if a map $y \mapsto f(y)$ of the line into itself is continuous, then (1) if the functional equation has a periodic solution of period k then it also has periodic solutions of any period following k in (*), and (2) if the equation has no periodic solution with period k then it has no periodic solutions of any period preceding k in (*).

The author is grateful to Yu. M. Berezanskii and Yu. A. Mitropol'skii for their helpful advice on the manuscript of this work.

References

1. Sharkovsky, A.N.: Ukrain. Math. J. **12**(4), 1960.
2. Sharkovsky, A.N.: Dokl. Akad. Nauk SSSR **139**(5), 1961.

Submitted 22.03 1962
      Kiev

Translated by J. Tolosa

Coexistence of cycles of a continuous map of the line into itself

A. N. Sharkovsky

S u m m a r y

The basic result of this investigation may be formulated as follows. Consider the set of natural numbers in which the following relationship is introduced: n_1 precedes n_2 ($n_1 \preceq n_2$) if for any continuous mapping of the real line into itself the existence of a cycle of period n_2 follows from the existence of a cycle of period n_1. The following theorem holds.

Theorem. The introduced relationship transforms the set of natural numbers into an ordered set, ordered in the following way:

$$3 \prec 5 \prec 7 \prec 9 \prec 11 \prec \cdots \prec 3 \cdot 2 \prec 5 \cdot 2 \prec \cdots \prec 3 \cdot 2^2 \prec 5 \cdot 2^2 \prec \cdots \prec 2^3 \prec 2^2 \prec 2 \prec 1.$$

Printed in the United States
by Baker & Taylor Publisher Services